WHEN EINSTEIN
WENT TO ROSWELL

WHEN EINSTEIN WENT TO ROSWELL

UFOs: The Conquest of Gravity and Space

PETER STRASSBERG, M.D.

Full Court Press
Englewood Cliffs, New Jersey

First Edition

Published in the United States of America
by Full Court Press, 601 Palisade Avenue,
Englewood Cliffs, NJ 07632
fullcourtpress.com

ISBN 978-1-946989-97-0
Library of Congress Control No. 2021914190

Editing and book design by Barry Sheinkopf
Graphic art by Richard Donatone (donatonedesign.com)

TO MY MANY PATIENTS
from whom I have learned so much

TABLE OF CONTENTS

———■———

INTRODUCTION

W E ARE NOT ALONE. The universe is not what we believe it to be. In 1947, Albert Einstein, along with several other equally prominent scientists, was shown the Roswell, New Mexico, UFO and its dead extraterrestrial occupants. The realization of visits by otherworldly beings led to a profound unsettling of his fundamental beliefs.

The United States has, subsequently—from this and other UFOs—back-engineered much of today's technology. Lasers, transistors, integrated circuits, fiber optics, infrared vision, Kevlar, maglev trains, and even the binary code for computers were due to seeds planted by advanced civilizations.

Gravity, best described by Isaac Newton, then further refined by Albert Einstein, is today, due to our contacts with advanced extraterrestrial cultures, fully understood. Inertia, with its similarity to gravity, can now be logically explained.

The supposed Big Bang theory beloved by most scientists, is simply wrong. Trying to prop it up with *ad hoc* inflation and dark energy has led modern physics down hopeless rabbit holes. The universe can be much more easily understood as the three-dimensional "bent" surface covering the fourth dimension of time. Hubble's redshift and the cosmic microwave background can, then, both be easily explained.

The world is granular. It has a smallest size. This can be inferred from Zeno's original paradox (concerning the impossibility of motion) and was proven almost 2,500 years later by the renowned physicist Max Planck. The smallest possible size is an "object" with Planck length dimensions of 10^{-35} meters —100 million, trillion times smaller than the tiny proton. Yet although miniscule in size, its mass and potential energy become the greatest possible.

Once the universe is understood as a *surface*, but in *three dimensions*, one can begin to visualize how it is maintained by each proton. Gravity can be seen as an emanation from the many protons (10^{80}) that constitute all that exists.

Finally, given that we are not alone, that we have been visited for eons, and that we have been genetically engineered from our ape-like ancestors, our current place as a space-faring culture, can be explored.

The final chapter of this book gives context to this history. Included in the Appendix are two remarkable governmental documents, reproduced in full, that were obtained via FOIA and other sources and are available on the internet. Assuming that they are accurate, they blow the lid off of 70-plus years of secrecy and subterfuge.

If you are interested in a quick perusal of the UFO material, just read the first and last chapters. You can then delve into the governmental information, as much or as little as you like. If you're interested, however, in the absurdity of the Big Bang, in understanding gravity and inertia, and seeing how a flying saucer works, I recommend the rest of the book. Enjoy.

Chapter 1
ROSWELL

ALBERT EINSTEIN WAS WRONG. He confided as much in a 1949 letter to his close friend Maurice Solovine: "You imagine that I look back on my life's work with calm satisfaction . . .[however,] there is not a single concept of which I am convinced that it will stand firm, and I feel uncertain whether I am in general on the right track."[1]

Why the misgiving? Why the disillusionment? We know that, just thirty years before, Einstein had been quite confident of the soundness of his ideas. When asked about the 1919 solar eclipse findings (bending of light by the Sun) that had confirmed his theory, and the possibility that they could have, instead, disproved it, he had replied, only somewhat facetiously, "Then I would have been sorry for the dear Lord, the theory is correct."[2]

---

[1]Manning and Manewich, (2019), *Hidden Energy*. Victoria, BC: Friesen Press, p.95.
[2]Walter Isaacson, (2008), *Einstein His Life and Universe*. New York, NY: Simon & Schuster Paperbacks, p. 259.

Nikola Tesla, Einstein's equally brilliant contemporary, had, on the other hand, always felt that Einstein's concepts were unsatisfactory. He had likened them to a beggar dressed in the purple robes of royalty, masquerading as a king. He felt that, although the mathematics may have been valid (the purple robes of royalty), they still covered a fallacy (the beggar) and did not reveal the underlying truth (the king).

What made Einstein, now, second-guess his conclusions? What facts or incidents intruded into Einstein's world-view? What so upset his equilibrium, causing his profound self-doubts?

The letter to his friend was written in March 1949. Less than two years earlier, in September, 1947, Einstein, along with several other equally prominent scientists (J. Robert Oppenheimer, the head of the atomic bomb "Manhattan Project"; Theodore von Karman, the leading American physicist associated with rockets and supersonic flight; and James Doolittle, the air force general famous for his 1942 bombing raid of Tokyo and a respected aeronautical engineer) were asked to help with interpreting the extraordinary findings of the Roswell, New Mexico UFO incident.[3] They were shown the damaged "space craft," with its dead aliens, that had crashed in July, 1947. Their scientific judgments were:

The findings were extraterrestrial, beyond the technology of United States science or even World War II German rocketry.

---

[3]*See Appendix B: White Hot Report*

—

Interplanetary space travel was possible.

Our solar system was not unique—there likely was intelligent life on other planets.

Those other cultures may have developed faster or be much older than ours.

Human origins may not be limited to our planet but may be found among other similar solar systems.

The laws of physics and genetics might be more complex than previously thought.

The final consensus was that: "Given the existing political climate in the U.S. and the unstable conditions in Europe [just two years after the war]. . .that if the Administration went public with the information as found in this report now, the results would be damaging, *even fatal* to the world political structure as it now exists" (*emphasis added*).

It was decided, therefore, to keep these findings secret. A panic, similar to the one created by Orson Welles in his 1938 *War of the Worlds* broadcast, was very much feared. Moreover, the findings were considered of great military importance; they did not want the Soviets to get even the slightest inkling of them.

As a result, the Roswell incident with its "other-worldly" findings was camouflaged under the rubric of a weather balloon mishap. Einstein, however, shown the actual evidence, knew the astounding truth: There *were* advanced extra-terrestrials visiting Earth. They most likely had come from other

solar systems surrounding distant stars, traveling vast ex-
panses at superluminal (faster than light) speeds. His theories,
based on the sanctity of light's velocity as the absolute limit
in the universe were, therefore, crumbling. This knowledge
percolated through his psyche and led to a deep and profound
unsettling of his world-view.

Chapter 2

EINSTEIN

EINSTEIN'S CONCEPT OF GRAVITY, his theory of relativity, has had a profound effect on science. Gravity, he felt, was the result of a curvature or warping of space caused by massive celestial bodies. The solar eclipse of 1919 proved this. It showed how light waves, always traveling in a straight line (the shortest distance between two points), were bent by the Sun.

Prior to Einstein, most felt that the concept of gravity was well understood. It had been *fully explained* by Isaac Newton's theories. Newton had shown how all objects—all things composed of matter—attracted one another. This force of attraction was *instantaneous* and, although greatly diminished by huge distances, was always present.

Einstein's theory was a further refinement of Newton's world-view. Einstein had based the theory (of relativity) on the concept of *equivalence*. He felt that gravity could be understood simply as acceleration. There was no "real" force involved. Therefore, if a rocket far away in space suddenly accelerated, that motion would be felt as similar to a gravitational force. A voyager in that rocket would be pulled in the opposite direction. If freely floating, that individual would suddenly fall to the floor. This force would be exactly analogous to what is felt on Earth. It would be thought of as a gravitational tug, a force of attraction. The two would be *equivalent.*

NEWTON

Newton's concept of gravity, the force contained in all matter, pulling on all other objects with its strength diminishing rapidly over distance, was based on what he considered the fundamental laws of physics. Succinctly, they are: *A body in motion stays in motion, unless struck by an outside force; and every action has an equal and opposite reaction.***

---

*Einstein believed that there was no "force" of gravity; there simply was an accelerating surface—either the rocket or the Earth—pushing up against a non-accelerating or "freely floating" body. Thus, if someone on Earth falls from a height, that person, while falling, is simply "freely floating." It is the Earth which is accelerating and "rising up" to hit that person. It becomes similar to the example of an astronaut dropping an object while in an accelerating spacecraft. That object, once dropped, is also freely floating; but since the rocket is accelerating, its floor rises up and hits the object.

**Newton's laws can be written as follows: A body moving at a constant velocity, or at rest, will continue unchanged until disturbed by an outside force. This outside force is some agent that directly causes a change, that is, speeds up or slows down that body's motion. When so struck, that body will impart an equal but opposite force upon the striking agent.

—
9

These essential ideas are the foundation of physics. They explain how the world works. Force must be *directly* applied to cause its effect. Newton, however, was perturbed by his own explanation of gravity's action. He could not understand how a force could act at a great distance on objects, without direct contact. It made no sense, given these fundamental concepts. Einstein, later, when trying to come to terms with quantum theory, and *action at a distance*, had the same misgivings. He called those findings "spooky," or unreal. He looked in vain for some deeper truth that could explain the bizarre quantum realm of tiny particles and waves.

We, therefore, have these two pillars of science, Newton and Einstein, both explaining gravity as either a force or acceleration, but both unwilling to accept the possibility of action without direct contact. Newton never fully accepted his own ideas; Einstein used the geometry of bent space to allow for gravity. But both fell short of a more fundamental understanding.

Chapter 3

ZENO'S PARADOX
AND GRANULARITY

WHILE TRYING TO SHOW THAT MOTION was impossible, Zeno, a Greek philosopher, some 2,500 years ago, described his famous paradox. Before someone could travel any distance, let us say 100 yards, that individual had to go one-half of the way—50 yards. But before going that half distance, that same person had to move one-half of that, and so on-and-on, *ad infinitum*. Hence, before starting on any journey, one had to take an infinite number of first steps. Since it is impossible to do anything an infinite number of times, all motion was illusory. The world was really static.

Of course, Zeno was wrong. The world is *not* static; it is in *constant* motion. However, although Zeno's world-view is inaccurate, his paradox is logically correct. How can we solve this?

The best explanation is that space cannot *be* infinitely divided—it cannot be cut into ever-smaller pieces. There has to be a least size to things. The world has to be *granular*; it cannot be completely smooth and diffuse—forever divisible. Max Planck, an important early twentieth-century physicist—the father of quantum theory—showed that there was a smallest possible size to all things, which subsequently was memorialized as the *Planck Scale.*

The shortest possible length is the Planck length (1.6×10^{-35} meters). The least feasible period of time is the Planck moment; the time it takes light to travel one Planck length (5.4×10^{-44} seconds). Finally, the smallest three-dimensional thing would be an object—a "cube" or "sphere"—of Planck length size. However, although it would be the tiniest conceivable object, its mass would be enormous; it would be almost 100 million, trillion (10^{20}) times that of the much larger, yet still miniscule, proton.

Given the reality of a granular world, we can begin to get a glimmer of a solution to the problem of gravity—how a force can act on something without any apparent physical contact.

AETHER

To further understand this we have to consider the concept of the "aether." Newton held that there had to be a "substance" through which the force of gravity worked. In a letter to the Irish philosopher, the Bishop, George Bentley, he stated:

"That gravity should be innate and essential to matter, so that one body may act upon another at a distance through a vacuum, without the mediation of anything else, by and through which their action and force may be conveyed from one to another, is to me *so great an absurdity*, that I believe no man who has in philosophical matters a competent faculty of thinking can ever fall into it. *Gravity must be caused by an agent* acting constantly according to certain laws; but whether this agent be material or immaterial I have left to the consideration of my readers"* (*emphasis added*).

Einstein also believed in the concept of something *permeating* space. In a 1920 speech, titled *Aether in the Theory of Relativity*, he stated:

"Recapitulating, we may say that according to the general theory of relativity space is endowed with physical qualities; in this sense, therefore, there exists an Aether. According to the general theory of relativity *space without Aether is unthinkable*; for in such space there not only would be no propagation of light, but also possibility of existence for standards of space and time (measuring-rods and clocks), nor therefore any space-time intervals in the physical sense. But this Aether may not be thought of as endowed with the quality characteristic of ponderable media, as consisting of parts which may be tracked through time. The idea of motion may not be applied to it"** (*emphasis added*).

---

* *Great Books of the Western World*, Vol. 45, Encyclopedia Britannica Inc., 1988, p.832.
** Manning and Manewich (2019), *Hidden Energy*, Victoria, BC: FriesenPress, p. 95.

—
13

Tesla, as well, felt very strongly that there had to be "something" for the transmission of force. An action could not cause an effect when there was nothing intervening. Concerning Einstein's concept of curved, empty space he stated:

"To say that in the presence of large bodies space becomes curved is equivalent to stating that *something can act upon nothing.* I, for one, refuse to subscribe to such a view"* (*emphasis added*).

Finally, Albert Michelson, who with Edward Morley performed the experiment in the 1880s that supposedly debunked the aether concept, later, in 1923, conceded that "Einstein's relativity and an aether *can* be compatible"** (*emphasis in the original*).

Thus we have a granular universe permeated with some unknown substance—Aether—through which light and gravity act. Could the smallest possible material, the Planck volume or "parton," be that substance? If so, can we find a common basis for electromagnetism and gravity?

Ibid., p. 94.
**Ibid.*, p. 93.

Chapter 4

THE FOURTH DIRECTION

I F THE UNIVERSE IS GRANULAR, and the smallest "particle" of space is the parton, what else might exist? Does space have a "background," or can it be its own milieu? Can particulate three-dimensional space exist in "nothing?"

It would seem, logically, that there has to be "something" else.

Our world is three-dimensional. There are three directions to all things: up-down, side-to-side, and front-to-back. But what if there was another direction? What if there was a *fourth* direction?

THE LOGIC OF A HIGHER DIMENSION

From a purely theoretical standpoint, if some universe could exist as one-dimensional, it would of necessity have to be shown as a straight line, but on a two-dimensional surface. Consider the following:

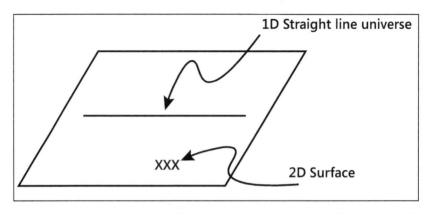

In a similar manner, if a world could be *two*-dimensional, then it would have to be placed in a *three*-dimensional volume. As follows:

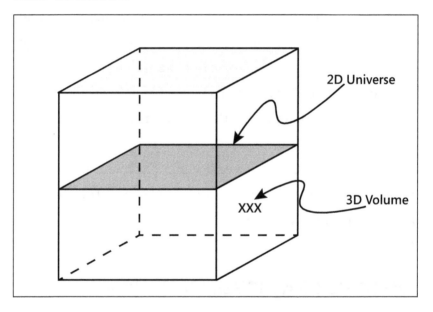

However, although we can in theory consider other, lower-dimensional worlds, in reality only our three-dimensional universe exists. But just as a straight line must be placed on a plane, and a flat surface within some volume, our three-dimensional world *also* has to reside within the confines of a higher dimension. It *too* demands the next dimension; but where is it to be found?

The next, or fourth, dimension would have to be at 90^0—orthogonal, or at a right angle—to the other three. But that does not seem possible, as *all* space is occupied by the knowable three. Hence, the fourth direction would have to be "within" the other three. If we draw it:

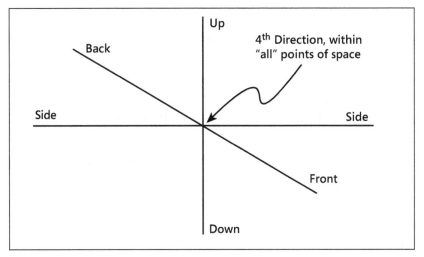

But if the fourth direction is within every point of space, and we also know that each point cannot be smaller than a parton or Planck volume, then the fourth dimension must exist *within* each parton, each indivisible granule of space. It also, logically, must be *between* each granule and the next, in that incomprehensible background upon which our three-dimensional world resides.

Chapter 5

BIG BANG

THE CURRENT SCIENTIFIC BELIEF IS THAT the universe started with a "Big Bang." This theory dates back almost 100 years, to a physicist who was also a cleric, Georges Lemaître. According to Einstein's general theory of relativity, the universe should be either expanding or contracting. Therefore, if it is expanding, by going back in time there logically had to have been an initial event. Lemaître felt that a starting point would have been a primordial "atom" containing all that currently exists. It would have "exploded," leading to today's world. He presented this concept to Einstein, who felt that, though the math was reasonable, the physics was not. Einstein had, instead, "fudged" his own formulas with a "cosmological constant." He did it to allow for a stable, unchanging universe —the accepted paradigm of that era.

Later, in the early 1930s, when Einstein toured the United States, he visited the celebrated astronomer Edwin Hubble and was shown his redshift data. Hubble had interpreted the "redshift," or widening of light waves with increasing distance, as evidence of an expanding universe. Einstein therefore realized that his original formula allowing for expansion had been correct. Lemaître's theory made sense, and his own cosmological constant had been a foolish error. In fact, he called it the *greatest mistake* of his life.

However, Lemaître's ideas still did not gain much traction. Hubble had miscalculated the rate of expansion and had estimated the age of the universe at about two billion years. Scientists knew this was not possible, since they had already determined the Earth to be at least twice as old. So the concept of an "exploding" primordial atom remained obscure.

The redshift findings nevertheless led others to consider competing theories of cosmic origin. The most influential one, called the "Steady State theory" and espoused by Fred Hoyle, a renowned astrophysicist, accepted the expansion of the universe but did away with the initial cataclysm. The theory called for a continual formation of hydrogen atoms as the universe grew larger. In fact, Hoyle felt that Lemaître's theory was not only improbable but even downright silly, deriding it as a "big bang." Later, however, as Lemaître's ideas gained mainstream credence, that moniker grew to become the accepted name.

COSMIC MICROWAVE BACKGROUND

These competing theories continued until the mid-1960s, when two scientists at Bell Labs in New Jersey, using their radio telescope, found cosmic background "noise" that could not be readily explained. It was microwave static that was coming in equal intensities from all parts of the universe. Other theorists had concluded that, if there had been an initial cataclysm, the infrared photons spewed out from that "explosion" would still be present and, since the universe had expanded about 1,000 times, they would now be 1,000 times as large. The initial wavelengths of these photons would have been about 1/1,000,000th of a meter (infrared dimensions), but now, given expansion, they would have grown to about 1/1,000th of a meter (microwave range). Hence, when this flood of microwave radiation was found (labeled the cosmic microwave background, or CMB), the current acceptance of the Big Bang truly began.

Although CMB cemented the current belief, there still were significant problems with the Big Bang theory. The universe was known to be "homogeneous," essentially the same in all directions. Why, given a "messy" initial cataclysm, was the world now so "smooth"? Also, the energy level of CMB was the same regardless of the direction in which one looked. Why would areas that could not have been in contact now exhibit precisely the same energy signature?

INFLATION

These questions were apparently answered by Alan Guth, an astrophysicist, who, in the late 1970s, theorized that there had been a very early, extremely rapid, "inflation" following the initial Big Bang. It led to a smooth universe just as any crinkly balloon would when blown up. Furthermore, prior to inflation, all parts of the initial explosion were close enough together to have possessed the same energy characteristics that subsequently were maintained even when separated by the tremendous ensuing enlargement.

No one could give a good reason for inflation, but the concept fit and answered the concerns that had been raised. The Big Bang premise, therefore, supported by CMB and buttressed by inflation, became the scientific community's accepted belief.

Later, in the waning years of the twentieth century, another question arose. Given that the universe was expanding, and that gravity—a force of contraction— was the only known force on a large scale, could this expansion now be slowing or even reversing?

DARK ENERGY

To answer this concern, a new determinant of distance was required. It was known that a certain class of supernovas—type 1a—could be used as a measuring rod for distance. These extremely luminous 1a events would always shine with the same brightness; thus, the dimmer they appeared, the farther away they were. In fact, one could accurately gauge distance using these "standard candles."

To the surprise of the researchers, when the study of 1a supernovas was undertaken, it was found that they appeared to be farther away than the theory had assumed. This extra distance could not be explained simply by an initial Big Bang with subsequent inflation. Instead, another energy source was required. Since none was known, this source was termed "unknown" or "dark" energy.

Today, most scientists believe that the Big Bang led to our present world. Anchored by CMB and reinforced with inflation, the universe exists in its present state. Gravity does not lead to contraction, as the cosmos is filled with a form of dark energy which maintains and increases its expansion. This is the accepted paradigm.

But is it correct?

I will show that the world is not constructed as present-day theorists believe. To accomplish this, I will use the previously noted concept of the fourth dimension. Given that "imaginary" direction, I will describe a more fundamental reality.

Chapter 6

THE REDSHIFT
AND A HIGHER DIMENSION

THE INITIAL RATIONALE FOR THE BIG BANG theory was Einstein's general theory of relativity. It described an unstable universe whose size was either increasing or decreasing. Lemaître came up with the concept of expansion after an explosive start; and Hubble, using redshift data, showed that these ideas were reasonable.

Hubble felt that the change in the wavelength of light, the redshift or lengthening of light waves coming from distant galaxies, was due to a Doppler shift. When an ambulance, its siren blaring, approaches, the pitch of that sound rises (its wavelength shortens). As it departs, that pitch falls (its wavelength expands). The faster the motion, the greater is the change. This change in wavelength is an excellent example of the Doppler shift.

Astronomers had known, prior to Hubble, that stars approaching us shifted their light toward the blue or shorter spectrum, and those moving away toward the red or longer spectrum. Hubble, with the use of the world's largest (100-inch diameter) Mount Wilson telescope saw, not only that all distant galaxies shifted their light toward the longer or red spectrum, but that the farther away, the greater was this shift. He surmised that all galaxies were moving away, the most distant the fastest. He had no reason to consider any other underlying factor.

However, if a fourth direction is posited, we can show an entirely different concept.

It will not require *ad hoc* inflation, or dark energy, and it can just as easily explain CMB. It will employ just one variable—the fourth dimension—and will therefore be much less complicated than the muddled Big Bang with its various patched-together corrections. As Occam's razor implies, simpler is better. The less complex, the more likely something will be right. Finally, using Einstein's special theory of relativity, we will demonstrate the Big Bang's fatal flaw.

ONE-DIMENSIONAL WORLD

Let us first assume that the world is one-dimensional. Of course, it is not and cannot be; however, if we use this analogy, it will greatly help to simplify our explanation. I will show how, although a one-dimensional universe is *physically* impossible, it is *mathematically* useful.

If the world is one-dimensional, a straight line, its inhabitants are segments of that line. Let us also assume that this "straight-line world," drawn on a flat surface, is, in reality, a circle. To us, this is easy to comprehend. To its "straight" inhabitants, however, a curve is impossible to visualize. What those minute segments of that straight line world "see" instead is a shifting toward ever-greater lengths. If we draw their world, we get a circle that its inhabitants can only understand as a straight line. They cannot conceive of a "bend" in their one-dimensional space. That would entail a higher dimension, a completely unknown direction. Instead, they sense ever-increasing redshifts with distance.

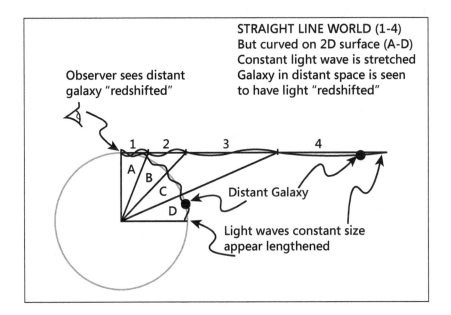

STRAIGHT LINE WORLD (1-4)
But curved on 2D surface (A-D)
Constant light wave is stretched
Galaxy in distant space is seen
to have light "redshifted"

Observer sees distant galaxy "redshifted"

Distant Galaxy

Light waves constant size appear lengthened

To a one-dimensional observer, a galaxy in deep space has its visible light waves lengthened or redshifted. This, however, is strictly a distortion due to a higher dimension. It is what we see, for example, in an ordinary map of the world—a Mercator projection. On that two-dimensional representation (of our three-dimensional globe), land toward the poles gets ever more stretched and distorted. Antarctica takes up the entire southern portion of the map.

To us, visualizing their world from a higher plane, distances on the circle (A-D) are equal; but to the one-dimensional observer, these distances (1-4) are constantly increasing. Light waves, or any other constant metric, will exhibit this continual lengthening, or redshifting. Therefore to our one-dimensional inhabitants there is a redshift increase with distance. Although our world is three-dimensional, the exact same thing happens.*

Hubble's findings are not due to a Doppler shift. There is no rapid movement of galaxies away from us in all directions. We are not part of an expanding universe. They, and we, are all part of a three-dimensional surface covering a fourth-dimensional "entity."

---

*In mathematics, a round object of any dimension is called a "ball," its surface a "sphere." The surface is always one dimension less than what it covers. Thus a baseball, for example, has a leather surface that would mathematically be two-dimensional (2-sphere) covering a three-dimensional interior (3-ball). All spheres, starting with a common circle (1-sphere) and including our universe ("hypersphere" or 3-sphere), can mathematically be expressed as the product of the circumference of a circle ($2\pi r$) and the interior of a sphere two dimensions less. Thus, our highly complex 3-sphere universe is the product of a circle's circumference and its two-dimensional smaller interior ($2\pi r \times \pi r^2$ or $2\pi^2 r^3$). The actual calculations are not important; what they signify is that all spheres, including our universe, can, from a mathematical standpoint, be considered simple circles.

Chapter 7

CMB—THE EDGE OF IT ALL

A LTHOUGH THE REDSHIFT WAS THE INITIAL PROOF of Lemaître's conjecture, the real acceptance came with the discovery of the cosmic microwave background. It was believed to be due to the original infrared photons of the Big Bang (1/1,000,000th of a meter) expanding with the universe 1,000 times as it aged, and becoming microwaves (1/1,000th of a meter). Fred Hoyle's Steady State theory could not account for this finding, so it faded, and the Big Bang became the established belief.

However, if we accept that we are the surface of a hypersphere composed of granular space particles (Planck volumes or partons), then the farthest out we can see are the most distant or very last partons. They would be 13.8 billion light-years away, the time it has taken light from the *beginning* to reach us today. Since light is the measuring-rod of our world, 13.8 billion years of its travel is equivalent to about 10^{26}, or 100 trillion, trillion meters.

We have shown that the farther away an object is on our 3-sphere curve, the longer or more stretched it appears. Thus, the most distant object, the last parton of visible space, would have the maximum stretch possible. Astronomers use the term "z" for this elongation; and the z factor for the most distant parton can be shown to be 10^{32} or 100 million, trillion, trillion times what it would appear if close up.[*]

But if something 10^{-35} m (the approximate size of a parton), was expanded 10^{32} times, it would appear to us as if it were 10^{-3} m (10^{-35} m x 10^{32} = 10^{-3} m). It would seem, that is, as if space were composed of microwaves, 1/1,000th of a meter in size. This is why, no matter where one looks, there is an extensive blanket of microwaves (CMB) radiating toward the viewer. Although they come from all directions and from distances too great to have ever been in contact, they are, nevertheless, all the same.

Finally, although CMB appears uniform, there are irregularities of about one part in 100,000. These are theorized to be due to quantum fluctuations that have expanded with inflation into today's galaxies and clusters. However, they can be more accurately explained as already existent galaxies and their clusters seen fully stretched at the edge of the world.[**]

---

[*] 10^{32} is the stretch factor (z) of a light wave after which that wave would be greater than the size of the universe and therefore invisible. It is based on an average visible light wave of about 500×10^{-9} m or (5×10^{-7} m) and, when compared to the radius of the universe (10^{26} m), is found to be 2×10^{32} (10^{26} m/5×10^{-7} m) or, if rounded down, 10^{32} times smaller. Thus a light wave is $1/10^{32}$ the size of the universe and, if stretched slightly more than 10^{32} times, becomes bigger than the world.

The concept of a higher-dimensional stretch answers the finding of CMB and gives a simpler reason for both its homogeneity and its imperfections (existent galaxies and their clusters). There is no need for an initial Big Bang with infrared photons that have expanded to microwave size. There is *certainly* no reason for an inflationary stage to smooth out this background. It simply is what exists at the last possible visible boundary.

We have shown that the redshift is more easily explained as a higher-dimensional curve, and that CMB is its boundary.

What about the concept of dark energy?

THE "DARK" WORLD

Dark energy was the *ad hoc* answer to something that seemed to be needed to counter, not only the contractile force of gravity, but also to explain the greater distances than foreseen by Big Bang theory. Type 1a supernovas were found to be farther away than expected, and this was best explained by some "push" of energy over and above the original Big Bang "explosion."

Dark energy also helped to explain why the universe was "flat." Scientists believed that, although it was expanding, it was doing so in the three directions of space in a *straight* line fashion. They felt that there was no curvature to the entirety,

---

** An average galaxy is about 10^5 (100,000) light-years in size, whereas the universe is approximately 10^{10} (13,800,000,000) light-years from any spot to its edge. Thus a galaxy is around 1/100,000th the size of the universe ($10^5/10^{10}$). Hence, galactic structures (essentially, swirling clouds of hydrogen atoms congealed into stars) in otherwise uniform space give rise to these irregularities. They take up 1/100,000th of the entire volume of the universe; therefore, they present as irregularities of 1 part in 100,000. As before, current theory has this backwards.

---

that it was simply an expanding three-dimensional entity—it was "flat," not bent. But they also knew that there was not enough energy or matter (Einstein's famous formula, $E = mc^2$, shows them to be the same) to maintain a flat world. In fact, given the totality of visible matter, they were missing a truly huge amount of mass; over twenty times as much was needed to allow for flatness. Visible substance accounted for less than 5 percent; 95 percent, or more, was still required.

Astronomers had previously theorized that, besides visible matter, there was an unseen vast quantity of invisible or "dark matter." It was over five times as massive as was all visible or known substance. It was what allowed for the rapid rotational movements of galaxies and their clusters. Calculations of these speeds had demanded more mass, and since none could be found, scientists had to allow for some mysterious "matter" to fill the need. However, even with this extensive unknown quantity, they still needed almost four times as much "substance" to flatten the world. (Around 70% was still missing, according to theoretical calculations).

Dark energy not only allowed for this vast energy source, but it was found to be just enough for the amount required. Scientists, hence, had a good reason to accept dark energy as real. Just like dark matter, it filled an important need; nonetheless, in a similar manner, it could not be explained. Therefore, given the fundamental belief that the universe was flat, scientists accepted that 95 percent of all that existed was unknown.

Dark energy, just like inflation, was an *ad hoc* solution to the problem brought on by an ardent and stubborn belief in Big Bang theory. I will show, in the next chapter, that the entire concept of dark energy is unneeded and that the Big Bang itself is unsupportable.

Chapter 8

ILLUSORY DARK ENERGY

THE WORLD IS THE THREE-DIMENSIONAL surface of an unknowable fourth dimension. The mathematics describing that shape leads to accurate cosmic distances. A 3-sphere curve explains the universe as it *is*; no dark or expansionary force is needed. The attempt to solidify an unreal theory (Big Bang) using an illusory force (dark energy) is a fool's errand. With fairly straightforward math we can show that, up until about 12 billion years ago, the Big Bang theory gives us distances that are *too close*. Thus, when 1a supernova data showed greater distances than expected, it was not because of "dark energy"; the results were merely correcting a mistaken theory. *(Please see Appendix A.)*

In the following table, the distances calculated by using a fourth-dimensional concept (column 3) are greater and more accurate than those calculated through standard Big Bang theory (column 2) for all stretch or z factors (column 1) until just less than 12 billion light-years. The difference in distance is what erroneously has been considered dark energy (column 4).

1	2	3	4
Stretch Factor ("Z")	Hubble/Expansion Dist. (10⁹ Lys)	4D Curve/Tangent Dist. (10⁹ Lys)	Dark Energy Dist. (10⁹ Lys)
0.1	1.5	4.6	+3.1
0.2	2.8	6.1	+3.3
0.4	4.6	8.0	+3.4
0.5	5.5	8.5	+3.0
0.8	7.7	9.7	+2.0
1.2	9.5	10.7	+1.2
2.0	11.5	11.6	+0.1
2.4	12.3	11.9	-0.4

This same information is presented as a graph; the supposed dark energy becomes the area between the two curves.

TIME DILATION AND BIG BANG'S DEMISE

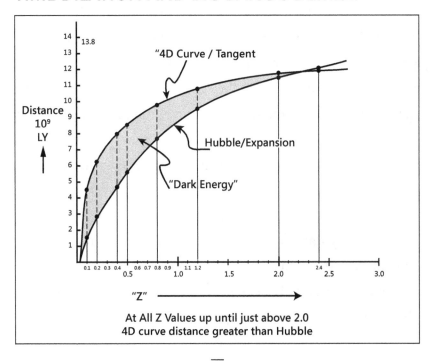

At All Z Values up until just above 2.0
4D curve distance greater than Hubble

I have shown that a more concise concept—a fourth-dimensional curve—answers all the concerns of Big Bang theory. It is simpler than unproven claims of inflation and dark energy. It gives the reason for CMB and its homogeneity. However, it does not *disprove* Big Bang theory. To do that, one needs to invoke Einstein's special theory of relativity.

Einstein showed that, as velocity increases, length shortens and time dilates. These finding are so minute at "normal" velocities that we almost never see them. Nevertheless, they start to manifest at *relativistic*, or "close- to-light" speeds. At about 85 percent of the speed of light, time doubles and length halves. Thus, an object would appear to age twice as slowly, but shrink to one-half of its length. Since the speed of light is 300,000 kilometers/second, 85 percent is about 250,000 k/s. Although this is an extreme velocity, if one believes in an expanding world, distant galaxies should be moving from us at that rate or more. But if they are, occurrences that take a known amount of time when at rest should take twice as long when moving that fast.

The 1a supernovas are just such events. They are not only exceptionally luminescent—visible at well over 10 billion light-years—but are very constant in duration. They take about twenty days from beginning to end. Given the theory of relativity, these 1a supernovas seen toward the edges of the universe, theoretically fleeing from us at relativistic velocities, should last, from our standpoint, forty days or more. However, when observed at these great distances, they still "flare-out" in the same twenty-day period.

Hence, either Einstein was wrong, which is extremely unlikely as his theory has been proved correct in *all* circumstances, or these supernovas are *not* departing from us at relativistic speeds. But if they *aren't*, then Big Bang theory is wrong. The world is *not* expanding. It is, instead, a three-dimensional surface bent toward an unknowable fourth direction.

I have therefore shown a more concise and likely explanation for what exists. There was no Big Bang. The world is not currently expanding. There was no inflation, and there is no need for dark energy.

A clearer understanding of gravity is now possible.

Chapter 9

SPHERES AND BALLS

THE WORLD IS A HYPERSPHERE, THE SURFACE of a fourth-dimensional "entity." Although it is thought to be flat, the redshift, misunderstood by most scientists, actually proves its curvature. As I said earlier, a round object in any dimension is mathematically called a "sphere." What it surrounds is termed a "ball." Therefore, in our three-dimensional world, a spherical object, a baseball, for example, theoretically has a two-dimensional surface and a three-dimensional interior. It is a 2-sphere surrounding a 3-ball.

Now, this is *strictly* mathematical. In reality, there are no two-dimensional surfaces. All things are three-dimensional. This book's pages have a thickness. The print on each page has a depth. However, in the mathematical realm, one can be more abstract. So, if we are considering a two-dimensional world, a round object that we commonly call a circle is now a 1-sphere. It consists of a one-dimensional line curving around a two-dimensional interior. That interior is a 2-ball.

In our day-to-day world, any regular sphere has an exterior surface and an interior volume. Our universe, however, surrounds an exotic fourth-dimensional core. In mathematical terms, it is a 3-sphere, and it encloses a 4-ball.

VISUALIZING HIGHER DIMENSIONS

If we were living in a two-dimensional world, on an entirely flat surface, what would a three-dimensional globe look like? We would not be able, as two-dimensional beings, to understand the concept of *height*; that dimension just would not exist. We would simply perceive the globe as a series of concentric circles, from the smallest in the center to ever enlarging ones as we looked toward its equator. Let us draw it:

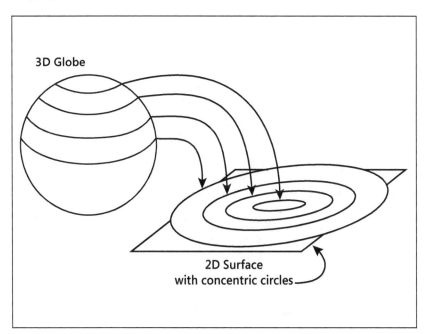

3D Globe

2D Surface
with concentric circles

We, however, live in a three-dimensional world. We cannot visualize the direction of a higher dimension. We call it "time" and can only assume that it is found within the other three, inside each point or granule of space. But if a fourth-dimensional object, a hypersphere, were to be described, we—in a fashion similar to that of the inhabitants of the two-dimensional world—would conceive of it as concentric "spheres," one inside the other. The spheres would extend from the smallest possible at the center to the largest at the periphery. Therefore, since we cannot comprehend a direction that does not exist, in order to draw a hypersphere, we show it as concentric spheres, one within another. Let us draw it:

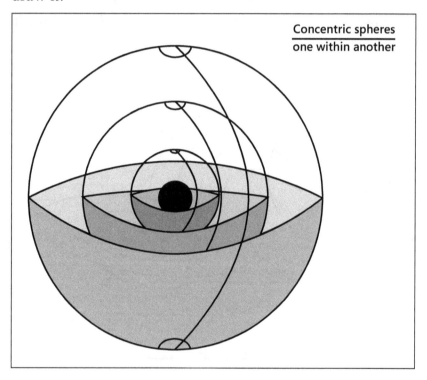

Concentric spheres
one within another

HYDROGEN ATOM

Our universe is made up of matter surrounded and filled with empty space. Although there are over a hundred separate elements and a vast multitude of combinations, hydrogen, by far, is the most abundant. We know that 90 percent of all atoms in existence are hydrogen atoms. They account for about 75 percent of the entire mass of the universe. They are found in the stars and in great nebulous clouds of gas and plasma.*

After hydrogen, the next most abundant element is helium. It makes up almost 25 percent of the remaining mass of the universe. This leaves only about 1 percent, or so, for all the others. Thus, to us, living on the Earth, although we are made of carbon, hydrogen, and oxygen with a smattering of many other elements, we, our planet, and all the innumerable planets like ours are, in total, but 1 percent of the mass of the universe.

Hydrogen is not simply the most abundant element; it is in many ways the *only* element. It is fused in the nuclear furnaces of the stars into helium, which in turn combines into ever-more complex configurations. Finally, it is thought that, in the cataclysmic explosions of supernovas, the more than

---

*Plasma is the fourth state of matter. When an element is cold enough, it is solid. As it heats up, it liquefies, and then, when sufficiently hot, it morphs into a gas. Finally, when highly energized, the proton-electron bond is so loosened that it becomes plasma. At this point, individual protons are no longer in direct contact with their surrounding electron clouds.

one hundred other elements are formed.

Thus hydrogen is really all that exists. It simply combines with itself and forms the rest of matter. So, in order to simplify our world, we can consider it to be composed *solely* of hydrogen atoms.

Chapter 10

TIME—THE IMAGINARY DISTANCE

I HAVE BEEN DESCRIBING A HYPERSPHERE. It consists of granular space (partons) and tangible matter (hydrogen atoms). The fourth dimension appears to be imaginary, but it really does exist. Every hydrogen atom can be shown to be at the same fourth-dimensional "distance" from what is considered the beginning of time.

Although this is difficult to visualize, using a lower-dimensional example can bring it into focus.

If we had a two-dimensional world (a flat plane) curved about a three-dimensional space (a common globe or 2-sphere), then every point on that surface would be the same distance away from the center of that sphere. If each point on our 2-sphere were a cognizant being, each would consider itself the center of that universe. Each could look back an equal distance to the beginning of time—the "Big Bang" of that world. Let me draw it:

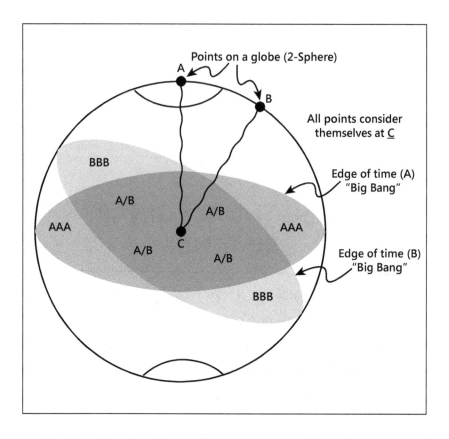

Each point (intelligent being) sees itself as the center of the universe (point C). Each is, at any instant, equally distant from the edge or—what could be considered that universe's beginning—from its Big Bang. The inhabitants cannot conceive of the direction of height. When told of this third dimension, they consider it time; and all are at any given moment equally distant (in time) from their Big Bang.

In our three-dimensional world, the curved surface of

an unknowable fourth dimension, our 3-sphere's distance from the center to the edge of time is the same for every hydrogen atom. To draw it we simply put all hydrogen atoms at the centers of equally sized spheres. The radius of those spheres is the time to the supposed Big Bang (13.8 billion years). Our 3-sphere universe, then, consists of a multitude of hydrogen atoms (10^{80}), each as the center of an ordinary sphere or globe (2-sphere), and each at the same central point—the same moment in time. *All are one in the fourth dimension*, as all are equidistant from the beginning.

Now, there was no actual beginning or Big Bang. Instead, there is a curve in three-dimensional space, reaching 90^0 at 13.8 billion light-years, beyond which nothing can be known. All that "exists" is within this limit. Although most consider it the beginning of time, it is merely the edge of the world beyond which no signal can extend.

SKETCHING A 3-SPHERE

Drawing a 3-sphere is difficult; nevertheless, it can be attempted in several ways. One is to take a three-dimensional cube of space, flatten it (still leaving it three-dimensional), and then "bend" it toward the imaginary or higher dimension. If we then place it on some incomprehensible "round" object, it becomes the cover of that entity. We now have a representation of a 3-sphere. Let me draw it (see this illustration on the next page):

—

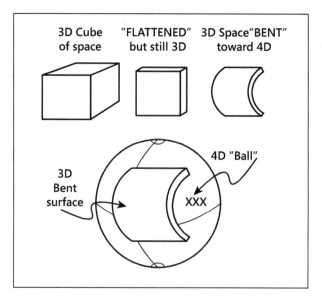

Another attempt involves cutting *through* a sphere. What we then get is a sphere of one lower dimension. Thus if we slice a 2-sphere by a plane, we get a 1-sphere:

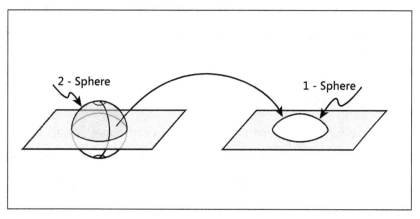

If we, therefore, were to transect a 3-sphere by some "wedge" of space, we are left with a normal globe or 2-sphere. The object that has been cut is our 3-sphere, and it resembles the following. Let me draw it:

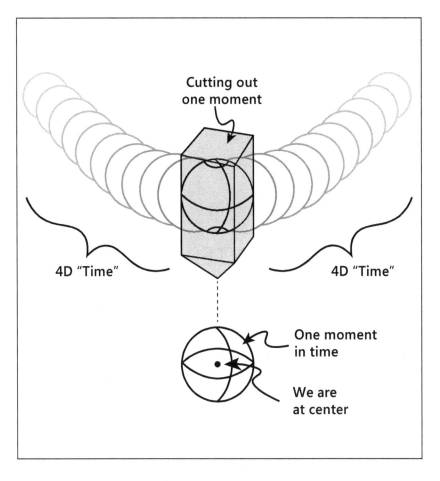

We have, therefore, drawn the equivalent of a hyper-sphere. It is a series of ordinary spheres, but situated in time—the fourth dimension. Each Planck instant, a new one comes into existence. Each one is a snapshot of our universe at that moment. Each has the same 10^{80} protons, and each proton, if sentient, would believe it was at the center of the universe.

Chapter 11

SPOOKY CENTER AND EDGE

LTHOUGH MOST SCIENTISTS BELIEVE that the world began 13.8 billion years ago in a cataclysmic Big Bang, if we consider the universe to be a 3-sphere and time a physical "distance," every particle becomes "equidistant" to that beginning. Furthermore, at any given moment the universe must be visualized as a three-dimensional sphere; thus its border represents its entire surface.

A proton, at 10^{-15} meters, is about $1/10^{41}$ the size of the entire universe (10^{26} meters). It takes almost 10^{41} protons to equal the radius of our world. (To simplify, we will use 10^{40}.) Its surface, then, is that quantity squared, or 10^{80} protons. (Interestingly, this is also the accepted value for the total number of protons in the world.)

The two-dimensional "surface" of our spherical universe is what encapsulates it. Although most, by far, of the world is empty space (100,000 trillion, trillion, trillion times as much as tangible matter), actual, physical things (we and all the planets and stars, and the great clouds of hydrogen gas and plasma) are formed of that surface. We are all at the same central point of a fourth-dimensional entity, equidistant in time to its beginning; however, we are also composed of its two-dimensional edge.

This dichotomy becomes the basis for "spooky" action at a distance, the quantum world's weirdness that so befuddled Einstein. For, if all tangible objects (all protons with their electron clouds) are really one, then strange interactions among *connected* particles that only *appear* to be separate make sense.

ALL ARE ONE

The concept—all are one—is hard to visualize in the "real" world. However, if we use a one-dimensional model, the concept comes into focus. In such a world we see that the redshift is really the result of a higher or second dimension. A curve into that direction leads to a continuous stretching of space as distance increases. Each point on the curve is a separate moment of time; hence, a separate universe is established for that tiny fraction of a second. All objects distant from that point are back in time—they are from the past.[*] Nevertheless, at any moment, all objects are at the same center in the universe. All are equally distant in time from its beginning. All are 13.8 billion years from the "supposed" Big Bang.

So how can all objects be at the same point yet separate? The answer lies in a higher dimension. Our one-dimensional universe, the circumference of a circle, covers the second dimension. That circle's center is equidistant from all moments in that curve.

The current moment—*here and now*—is at the "top." A tangent from that point corresponds to the present instant's *flat,* one-dimensional universe. It is a projection of all prior moments and positions in space. All distances continuously lengthen as one looks out into space and back into time, and the closer to 90^0, the greater their stretch. (As I have already shown, at just before 90^0 the stretch is the greatest possible—the basis for CMB.) However, all, at any moment, are equally distant from the center of the circle. All are the same 13.8 billion years from the beginning.

The center of that circle is a point of space—one parton. Since it represents all 10^{60} moments, it really embodies 10^{60} separate universes, each with its own central point. As a result, we have a total of 10^{60} partons, the volume of one proton.[**]

---

[*]*In our "real," three-dimensional world, everything around us is only sensed by the electromagnetic or light waves emanating from it; all that we "see" is from the past. If some object is 1 meter away, it takes light, travelling at 300 million meters per second (3×10^8 m/s), one 300 millionth of a second to reach us. So what we see has occurred in the past, 3 billionths of a second ago.*

[**]*A proton's diameter (10^{-15}m) is 10^{20} times as large as a parton's (10^{-35} m); as a result, it takes 10^{20} partons cubed, or $10^{60,}$ to equal the volume of one proton.*

Therefore the center of the higher dimension, from which all objects are equidistant, is a single proton. All that exists are projections of that proton. It establishes the universe for all 13.8 billion years. It allows for reality from 0^0 to 90^0 on a curve—from the present moment to the beginning of time. Let me diagram these concepts:

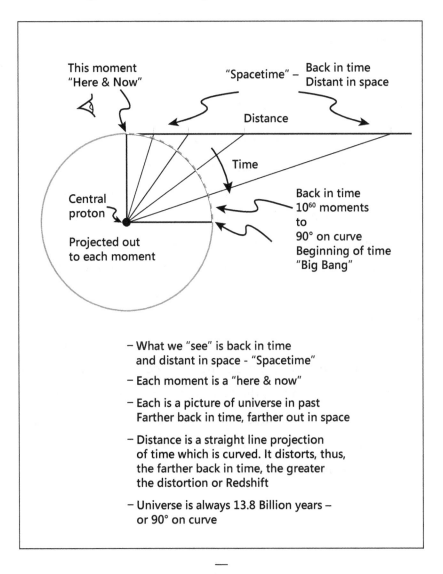

– What we "see" is back in time
  and distant in space - "Spacetime"

– Each moment is a "here & now"

– Each is a picture of universe in past
  Farther back in time, farther out in space

– Distance is a straight line projection
  of time which is curved. It distorts, thus,
  the farther back in time, the greater
  the distortion or Redshift

– Universe is always 13.8 Billion years –
  or 90° on curve

The universe is always 13.8 billion years old. It does not age. It continuously "moves" in time to the future. It reestablishes itself every Planck moment. Therefore, all protons are one—they are all based on a centrally located proton in a higher dimension. Spooky action at a distance, Einstein's great bugaboo, occurs because all objects really are the same. Each is projected in a different time; hence they appear, at any moment, as separate in space.

MOTION PICTURE WORLD

The world becomes the equivalent of a motion picture. The projector is the single, universal proton. Its film is the curved 3-sphere universe encircling a higher dimension. What we consist of are that bent surface projected onto a flat, three-dimensional screen. Each instant, a new universe unfolds. The minute changes occurring moment to moment in that higher dimension allow for motion and the transmission of light and energy.

Now, given that 10^{80} protons actually exist, the same quantity needed to make up the surface of our world, and given that this surface is the basis for all tangible objects, then everything of mass or substance, everything other than empty space, must essentially be formed from protons.

But what is a proton?

Chapter 12

PROTON

THE PROTON IS THE SMALLEST STABLE PARTICLE. Its diameter is 10^{-15} meters. Its mass is almost 2,000 times that of an electron. It is the core of the hydrogen atom. It is usually surrounded by an electron cloud that can be up to 100,000 times its size (10^{-10} meters). To get a proper perspective, if a proton were a tiny grain of sand placed on the fifty-yard line of a football field, the electron cloud would reach the goal posts.

The electron is held in place by the exceptionally powerful *electromagnetic force*. Its strength is 10^{39} times that of gravity (1,000 trillion, trillion, trillion times as strong). That is why a simple magnet stuck to a refrigerator stays in place; the entire world's gravity cannot pull it to the floor.

Yet although the electromagnetic strength is enormous, a more powerful force is seen within the proton—the *strong nuclear force*. It is about 10^{41} times as potent as gravity, or 100 times that of electromagnetism. This force holds a proton together and maintains the integrity of all nuclei.

The proton is believed to be a composite particle containing three smaller entities—*quarks*—held in place by the strong nuclear force. Although quarks "theoretically" fit today's concepts, their actual existence is quite tenuous. They are not found outside the proton and account for only 2 percent of its mass. Current theory allows for the remaining 98 percent to be found in the energy of equally theoretical *gluons*, exotic particles that, by their constant motion, keep or "glue" quarks together.

But no matter what actually exists within the proton, we do know that the force of attraction at its surface is the most intense in the universe. Even a neutron star, with its extreme gravitational attraction, exerts less force.

ENERGY OF TIME

Protons represent the two-dimensional surface encasing our world. They are strewn about space in the fourth dimension of time. Since they are enclosures, what do they enclose?

Our entire three-dimensional universe contains within itself a higher dimension. That fourth dimension is found hidden within the very granularity of empty space. Hence, what must be present within protons are the same granules, or partons, that engulf time.

We know that time extends back in our 3-sphere world 13.8 billion years. It, like space, is granular. Thus, time is composed of individual moments of existence, each lasting a mere 5.4×10^{-44} seconds. There are 8×10^{60} such instants. (For ease of use we will round down to 10^{60}.) Since each instant represents the entire, intact universe at that moment, then each instant must contain all of its energy.

The universe, as a surface, surrounds the *entire* fourth dimension of time, all 10^{60} prior instants. Hence the energy enclosed must be 10^{60} times as great, as can be measured. (Measurable energy is, essentially, found in the mass of its 10^{80} protons: $E = mc^2$.) Since each instant really holds 10^{60} times what is measurable, and there are 10^{60} instants, then there really is 10^{120} (one trillion, trillion, trillion, trillion, trillion, trillion, trillion, trillion, trillion, trillion) times more mysterious, *zero point energy* than is observable. This can also be shown in the huge mass of the partons (10^{20} times that of the much larger protons) that constitute our three-dimensional world. This is the enormous, untapped energy of empty space.[*]

---

[*]Each Planck volume or parton is about 10^{-35} meters in size. The universe, at 13.8 billion light-years, is just more than 10^{26} meters. Hence, around 10^{60} partons, if lined up side-to-side, would equal its radius. (Although the amount is closer to 10^{61}, I am rounding down to 10^{60} for ease of discussion.) Therefore, 10^{60} cubed (10^{180}) would more or less completely fill its volume. Since each parton contains an enormous amount of mass (almost 10^{20} times that of a proton), and since mass equals energy ($E=mc^2$), then the energy content of our universe is $10^{180} \times 10^{20}$ or 10^{200} times that of a single proton. There are estimated to be 10^{80} protons in our world; and the proton is, by far, the most significant source of energy in the observable universe. Consequently, the untapped or unknown energy, often called "zero point energy" (the energy present at absolute zero when there is no motion or measurable energy), is $10^{200}/10^{80}$ or 10^{120} times what is known. (Although there are an almost equal number of neutrons, with similar mass, doubling a number as great as 10^{80} makes no appreciable difference in these calculations.)

Protons are the two-dimensional "edges" of our three-dimensional world, which in turn covers a fourth dimension of time. Although *known* energy is found in the mass of its protons, the world itself encloses a much greater entity. It holds a teeming, seething, never quiescent energetic cauldron: the immense, untapped energy of the cosmos.

Chapter 13

NEUTRONS AND ANTIMATTER

T HE FORCE FOUND WITHIN THE NUCLEI of all atoms, the strong nuclear force, also maintains the neutron. Both neutrons and protons are present in every element other than hydrogen. What, then, is a neutron?

We know that a neutron can only exist within the confines of a nucleus. When freed, as in beta decay, that neutron rapidly (within 15 minutes) converts into a proton, an electron, and a rather unusual particle, an *antineutrino*. What has occurred upon this emergence?

If we consider the nucleus of an atom to be the extension of its protons (the two-dimensional cover to our world), then, as is seen in *all* surfaces, it must have two sides. One side faces inwardly to our world, the three dimensions of existence. The other side, however, faces outwardly to the unknown fourth dimension of time. The proton with its electron cloud, the hydrogen atom, is the three-dimensional side; the neutron, although exactly the same, faces the fourth-dimensional abyss.

Things on our side are known entities—protons, electrons and neutrinos—essentially hydrogen atoms and their many combinations. On the other side we have the same fundamental particles; however, they are considered the *antimatter* versions of what we know. Instead of a proton we have an antiproton, instead of an electron one sees an antielectron or positron, and finally, in place of a neutrino one gets an antineutrino.

These antimatter particles are similar to normal ones; they just appear "inside-out." They therefore have opposite charges: antiprotons are negative, and antielectrons are positive. This is the distortion of the fourth-dimensional "lens"; it reverses all aspects. Thus what was inside now is outside and charge, likewise, changes. Let me diagram it:

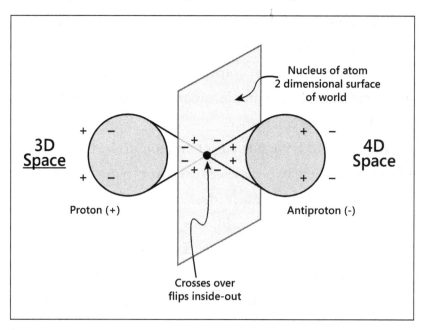

The best way to understand this is to consider positive charge a pull toward a center. Thus the proton is positive, as it is pulling things toward itself, the center of a hydrogen atom. The electron cloud, as it counters this pull, is negative. However, if seen in reverse, the antielectron is found as the center, the antiproton as the periphery. Thus, the antielectron is now what pulls and becomes positive. The antiproton, being pulled, becomes negative.

When we see something, it is from the outside looking in. If, for instance, we approach a football stadium, we first see its exterior walls; then, upon entering, we see its rows of seats. However, when we view the neutron we see it from the *inside looking out*. We see it in *reverse*. We are looking through the tiniest of openings—the nucleus of an atom. We are then using that aperture (the size of a proton) to gauge the rest of its structure, its electron and neutrino. We are within that grain of sand, situated in the center of the football field, gazing outward. However, although the stadium is much larger than the grain of sand, to us it can only exist within that grain. We cannot get closer, as all that exists in our three-dimensional world is that tiny grain. The enormous structure is confined deep within the smallest space.

The neutron, then, is a hydrogen atom presenting inside-out. The overall size of the neutron is the same as that of the proton; it is essentially the antiproton on the other side of our world's two-dimensional surface. Its much larger antielectron cloud is farther into that unknowable higher dimension; it is smaller and within the antiproton. The grain of sand and the football field have inverted; the grain becomes the outer aspect, the field its interior.

We cannot really view this higher dimension. It is within what exists. Hence, the electron cloud, as it is larger and deeper in that dimension, is more distant into that point of space. It is closer to an already infinitesimal center. Finally, the neutron also encloses a neutrino. It is much larger than the electron cloud but, due to the fourth-dimensional distortion, is farther within and exists as an antineutrino.

When the neutron escapes its nuclear core, it enters our third dimension. We see a standard proton and electron. It reverts to a hydrogen atom; it flips inside-out. The antineutrino, being much larger, has not fully escaped to the three-dimensional world and, consequently, remains in its antimatter form.

Chapter 14

PARTICLES AND FORCE

OUR WORLD IS COMPOSED OF SPHERES within spheres, from the smallest tangible one, the proton, to the largest possible, the universe itself. The force that contains these is seen as the strong nuclear force, the electromagnetic force, the weak nuclear force and, finally, gravity. We have protons within electron clouds further encased in evanescent neutrinos with nebulous energy spheres continuing to the very edge of the world.

Each proton is held together by the strong nuclear force. It is the centripetal force of attraction or, if written:

$$F = mv^2/r.$$

In this equation, force equals the mass of the orbiting particle multiplied by its velocity (squared), then divided by its distance from the center.

It is defined as the force pulling toward the center. As size increases, this force decreases. Therefore, the radius of the proton, 10^{-15} m, with a force of 10^{41} times gravity, is enclosing a particle 1/100 as large as is the electromagnetic force, 10^{39} times gravity. The weak nuclear force, with strength about 10^{34} times gravity's, should be enclosing a particle 10,000,000 times the size of a proton—the neutrino. Finally, the universe surrounds each and every proton, but is 10^{41} times as large.

The force of enclosure is the same for all "particles." This may be difficult to visualize; however, if we use a two-dimensional model, a flat surface that is curved into a sphere (a "beach ball" world, for example), it should become clearer. A proton in that world would be a small circle on that plastic ball, the electron cloud a bigger one surrounding it, the neutrino a much larger one, and the entire universe the spherical surface of the ball.

What is inside that sphere is equivalent to the "energy of time." It is compressed and held in place just as a beach ball contains pressurized air. A leak, no matter how small, would deflate the whole ball. Hence, every part of that surface, from the very smallest to the entirety, must be of equal strength.

The same holds for our world. But we just cannot *visualize* a hypersphere cover. It presents as spheres nested within spheres; but just as the beach ball requires equal strength on all parts of its plastic surface, so too does our world. The strength of enclosure is the centripetal force of each particle. Since centripetal force decreases with size, the smallest objects exert the most force. The strength is what we see as the energy inherent in mass ($E = mc^2$). Hence, the mass of the proton is about 2,000 times that of the electron and should be about 10,000,000 times that of the neutrino.* Let me diagram this:

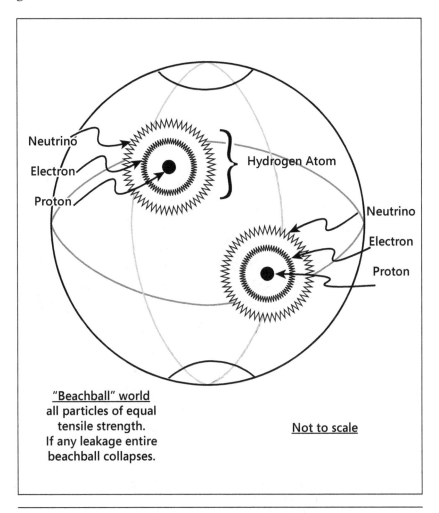

*The surrounding electron cloud extends outwardly from around 100 to 100,000 times the proton's diameter. Therefore, on "average," its size is about 2,000 times as great. Finally, although there are three different neutrinos, we will consider them as one. Their masses have never been well documented but are believed to be less than 1/1,000,000th that of the electron.

Therefore, our world, the hypersphere, visualized as spheres within spheres, has the same characteristic. Please see the following:

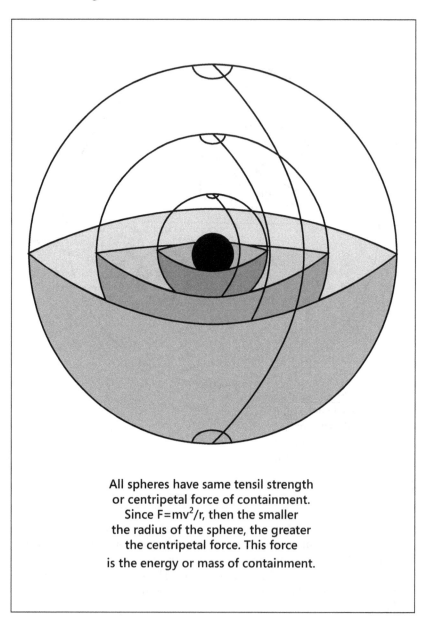

All spheres have same tensil strength
or centripetal force of containment.
Since $F=mv^2/r$, then the smaller
the radius of the sphere, the greater
the centripetal force. This force
is the energy or mass of containment.

NEUTRINO'S EASE OF TRANSIT

The mass of the neutrino (a particle that can travel through literally light-years of solid matter) is 1/10,000,000th that of the proton; therefore, it is 10,000,000 times as large. Because of wave-particle duality, its movement resembles the passage of a long-wavelength (infrared) photon through cosmic dust. The neutrino's size is much greater than the solid matrices encountered; hence, it is not interfered with during transit. This concept can be more easily understood using water waves. Let me draw it:

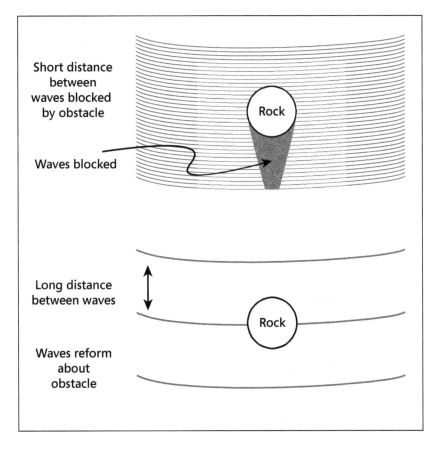

PROTON'S PERFECT PROPORTIONS

We can now see why the proton is the smallest tangible object in the universe. At 10^{-15} meters, it is 10^{20} times as large as a Planck volume, or parton (10^{-35} m). Hence, 10^{20} partons cubed (10^{60}) would fill its volume. Each parton has potentially 10^{20} times the energy of a proton; thus, 10^{60} times 10^{20} equals 10^{80} times the energy found in a proton's mass.

Most of the measurable energy of the universe is found in the mass of the 10^{80} protons of which all matter is composed. (I am leaving out an almost equal number of neutrons, each with similar mass, since doubling an amount that large makes no appreciable difference.) As a result, the *potential* or fourth-dimensional energy content of a proton equals the *real* or three-dimensional energy of the world.

Each proton's surface acts as the two-dimensional edge of our three dimensions of space (which covers the four dimensions of time). A *hypersphere* is a grouping of nested spheres, one within the other, from the smallest (particle) to the largest (entire universe). Since each sphere *must* have the same overall energy as they all do in total, the smallest possible one can only be the proton.

Every proton, with its volume of 10^{60} partons, each 10^{20} times as massive as that proton, encloses energy potentially equal to that of the entire universe. An object smaller than the proton, then, could not be a center to our world's hypersphere—it would enclose too little energy. An object greater in size also could not serve this function; it would have too much potential energy.

The proton is the exact size to be the core of our world's hypersphere. It is also the surface of our three dimensions of *reality* enclosing the *imaginary* fourth dimension. If there were no protons, there would be no boundary to our universe. It would not exist.

However, the protons' surfaces are not an all-encompassing and surrounding cover. They are not like the plastic surface of the make-believe two-dimensional "beach ball" world. The fourth dimension, as already noted, can only be understood as within our recognizable three. It resides within the partons of empty space. Protons, as the surface of our universe, hold within themselves only a very small portion of those partons; most by far exist outside of the protons' cores and make up the great expanse of our three dimensions. Nevertheless, the salient point is that *nothing* can be smaller or larger than the proton, yet still function as the stable, defining edge of our world.

Chapter 15

CENTRIPETAL FORCE

THE PROTON IS THE ESSENTIAL PARTICLE. It holds the world together. Its constant attraction is due to the centripetal force of its rotation; and this effect extends throughout the universe.

Each proton has within itself the *entire* energy of the universe. Each potentially has 10^{80} times as much as we can measure. Why, then, does this enormous energy source remain hidden?

Newton's basic laws of physics equate force with acceleration. A body remains stationary or in uniform motion unless an outside force is applied. Force, therefore, leads to change or acceleration; force is *directed* energy.

We know that the universe is filled with a tremendous amount of hidden or zero point energy. It is many trillions of times (10^{120}) what is observable. However, the universe's hidden energy is directionless. It merely gives rise to exotic particles that appear then quickly vanish. It exists within the partons of empty space. How, then, can it be put to use? How can it be given direction?

We have shown that the proton *must* exist. Each proton *is* the "center" of the universe. Each has all the "potential" energy within itself equal to the entire universe. However, for a proton to exist, it must contain and control that tremendous energy. This is found as centripetal rotation, as the strong nuclear force.

Centripetal force, therefore, is the energy of our world. It gives "direction" to the randomness of space. It is the "mass" within particles. It is found, mainly, within protons and neutrons but, also, to a lesser extent, in electron clouds. It is the basis for gravity, electricity, and magnetism.

Vast, empty space, although extraordinarily rich in energy, exerts no force. There is almost no direction to the constant churning of the "void." However, as noted, a minute amount of directed energy does exist. It is found in every particle's rotational movement. Although this is but a very small fraction, it is all that is assessable.

CENTRIFUGAL COUNTERFORCE

The electron's rotating cloud, with its balancing *centrifugal* force, counters the proton's attraction. Since the cloud moves as a single unit, it is only stable from its equator to around 60^0, north and south. Beyond those latitudes, its slower rotational speed cannot offset the proton's tug.*

—

Partons of space, therefore, are constantly pulled toward the proton, most being absorbed; however, some, just missing, are expelled outward, exerting a pressure or flux near the poles. This leakage inward and expulsion outward becomes the basis of both gravity and magnetism. In many ways, gravity appears similar to the van der Waal force that holds molecules together.**

ACCELERATION IS FORCE

Gravity, to Newton, was an instantaneous attraction, associated with *any* object of mass. However, as he told Bishop Bentley, he could not fathom how this attraction could occur without some direct contact. He did not know that the universe itself was particulate, made up of interconnected granules, each of immense but unfelt mass. He did nonetheless very clearly understand that force was always associated with

---

*Since the electron cloud moves as a single unit, its velocity (distance/time) decreases the shorter the distance covered in the same time period. It is comparable to velocity on the Earth's surface. The Earth's equator is about 25,000 miles in circumference; hence, the velocity at the equator is 25,000 miles/24 hours or a little more than 1,000 m/h. However, at the poles, we have a much smaller circumference of, let us say, 24 miles; hence, velocity there would be as slow as, or slower, than 1 m/h (24 m/24 h). On the solid Earth, this leads to a slight widening at the equator and flattening at the poles. However, in the tenuous electron cloud such diminished velocity does not allow for sufficient centrifugal counterforce to keep polar regions intact.

**Named for the Dutch physicist Johannes van der Waal, this is a very weak intermolecular force that is repulsive at very close distances (less than 0.4 nanometers), then attractive (from 0.4 to 0.6 nm), and, finally, as it is so inconsequential, becomes un-measurable (after 0.6 nm).

---

acceleration. Without acceleration or a change in velocity, nothing could begin moving or, if already in motion, alter its course.

Einstein also understood that force and acceleration were essentially the same. His concept of equivalence—equating a rocket's acceleration with a planet's gravitation—was due to this basic idea. However, he *also* did not take into account the granularity of empty space. We know that he believed in the existence of "aether," but he did not associate that nebulous entity with the reality of particulate space.

Therefore, employing the ideas of these great men but adding the reality of granular space, we can arrive at a fundamental understanding of gravity and, for that matter, all force. Gravity, then, becomes the residual of the attraction of each and every proton, one to another. It leads to an acceleration of partons and, hence, to a force based on that acceleration. *It is the acceleration of "empty" space.*

Chapter 16

DISTANCE AND FORCE

THE EARTH, AS A MASSIVE ROCK, CONTAINS approximately 10^{50} (100 trillion, trillion, trillion, trillion) protons. Each attracts all things to itself. Each proton is routinely surrounded by its electron cloud. The residual of this force of attraction is gravity. It is felt as a tug ever diminishing with distance. It extends to the very edge of the world. If a proton exists, so does its gravitational field. The field does not travel out from the proton at light speed. It, like the proton, merely *is*; and each proton with its extended energy field is the basis of our three-dimensional world.

Gravity is diminished by the square of distance. If twice as far, it is one-quarter as strong; at three times the distance, it has one-ninth the power. It is felt uniformly at every point equidistant from its origin. It is equivalent to the surface of a sphere. It always has the same overall strength; it merely spreads out to cover a larger surface.

While an object is falling, its velocity is increasing; it is accelerating. The force on it is growing exponentially greater. Since gravity has an equal effect on *all* things, the acceleration of a sizable body or an infinitesimal granule of space is the same. Empty space, therefore, is constantly being accelerated toward all objects; the greater the mass, the more forceful the effect.

INERTIA AND GRAVITY

Inertia is the jolt felt by a subway rider when a train begins to move. It is the reason for seatbelts when riding in a car. Inertia is present when there is a sudden start or stop.

Newton believed that inertia was a property inherent to all mass. Newton was wrong. Inertia is not an intrinsic part *of* matter. It is, instead, a passage *through* matter (of accelerated partons of space). They strike an object's electrons, leading to a sudden change in motion.

Gravity, too, can be understood as acceleration. It is based on the attraction of a huge grouping of protons (the Earth, for example) endowing "empty" space with the tangible mass that is always "potentially" there. Acceleration, brought on by gravity, allows for the emptiness of space to be felt.

Einstein saw this in his analogy of rocket flight with gravity. But he did not look for its underlying cause. Newton also understood this. But he did not conceive of empty space as having mass noticeable only upon its acceleration. Tesla comprehended this, as well. He felt that there had to be some intermediary; emptiness, or "nothing," could never be the basis for an action.

Gravity and inertia, therefore, are the same.

With gravity, we have space moving through an object. As the partons of space are accelerated, their mass is "awakened" and impacts that entity. Since the underlying pull is toward the multitude of protons that constitute some vast cosmic body, the force is constant and toward its center.

With inertia, the force is directed toward the object but opposite its motion (whether acceleration or deceleration). It is Newton's third, and most basic, law—*every action has an equal and opposite reaction.*[*]

Thus, a sudden jerk when a subway starts causes the "strap-hanger" to lurch in the opposite direction. The passenger's electrons have been accelerated (the train's abrupt start); it is not space but the *passenger* who accelerates and collides *with* empty space.

Therefore, the sudden acceleration causes the potentially massive partons of empty space, not normally sensed, to be-

---

[*]Newton's third law is essentially a statement about nature's symmetry. It is fundamental to physics and the study of "all" motion. If one, upon applying force to an object, did not receive a counterforce, then he or she would not "experience" the process of exerting a force. There would be no "mechanism" for any physical activity. (The counterforce that occurs arises in the material's intermolecular substrate, essentially in its orbiting electrons.) Therefore, if Newton's third law did not exist (for instance, if, when pushing against a wall, the wall did not resist with an equal counterforce), then, for all practical purposes, there would be no wall. The universe would be an entirely different place. Without counterforce there would, in essence, be nothing.

come a countervailing force. It is the *reaction* to an object's accelerating mass. It awakens the vast, latent mass hidden in the emptiness of space.*

SEAT BELTS AND UFOs

A seat belt in a car prevents injury at a sudden stop. It reels in the inertia of a decelerating passenger and tamps down the force of an abrupt change. It allows one to walk away unscathed. Any alteration of velocity, whether speeding up or slowing down, will cause the same counterforce of partons. In sudden stops, deceleration is from the front to the back; the equal but opposite force exerted by partons, therefore, is from the back to the front. Without appropriate encumbrances, an occupant strikes the windshield.

In a rocket suddenly accelerating, the opposite flow of particulate space pulls the astronaut to the floor. The space traveler feels a force as he or she accelerates through space. However, that person's acquaintances, on Earth, feel a similar force, as space, pulled by the many protons of the Earth, accelerates through them.

---

*This was shown by Stephen Hawking when he theorized that a black hole emits radiation. Its gravitational force continuously pulls on the emptiness (zero point energy) of space, and "virtual" particles become "real." Their existence is seen as radiation. Davies and Unruh later showed that any accelerating object likewise causes a discharge of radiation in front of its motion. These findings show that the acceleration of space through an object (black hole's gravitational attraction), or the acceleration of an object through space, both lead to a reactive force—radiation.

Although seat belts and other harnesses are appropriate for travel in normal conveyances, would they be needed in UFOs? We know that UFOs exist. The Navy now allows their pilots to discuss UFO encounters. However, we also know that they move at extraordinary speeds with astounding acceleration and deceleration. They can zip about in ways that should wreak havoc on their occupants, even if appropriately harnessed. How then is it possible for these otherworldly objects to function?

Chapter 17

HYDROGEN ATOM

THE UNITED STATES GOVERNMENT HAS BEEN actively involved in the research, design, and manufacture of antigravity vehicles since at least the mid-1950s. Around that time, numerous articles discussed this topic; since then, the subject has gone dark. Why the abrupt loss of scientific inquiry? Why the need for secrecy?

We now know that UFOs exist. The United States Navy acknowledges as much. Videos of their flights can even be seen on YouTube. They demonstrate an ability to negate both gravity and inertia. Why, now, are we being shown what, before, was taboo? Is our government finally trying to open its files?

Gravity is a force of attraction caused by vast concentrations of protons. Since protons are surrounded by fields of energy, and since hydrogen, with its single proton, is the fundamental constituent of all other elements, the following model of the hydrogen atom will show how it and, therefore, all things, interact with gravity and inertia.

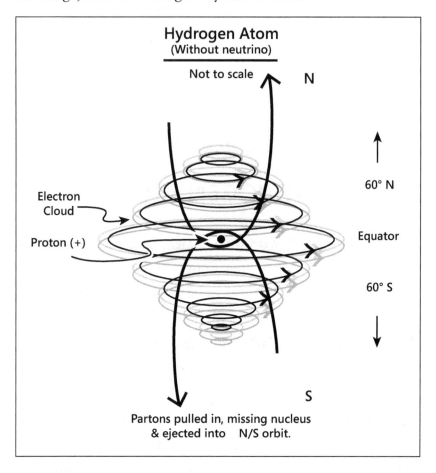

Hydrogen Atom
(Without neutrino)

Not to scale   N

60° N

Electron Cloud

Equator

Proton (+)

60° S

S

Partons pulled in, missing nucleus
& ejected into   N/S orbit.

COUNTERFORCE

As previously noted, the proton is much smaller than the hydrogen atom. It is about $1/100,000^{th}$ the size. It is our grain of sand in the center of a football field. Attracted to, and engulfing, the proton is the electron cloud, a stable, spinning vortex of space partons. Because of orbital rotation, the partons have mass and energy. They are the counterforce to the proton's pull.

This whirling maelstrom is most stable at its equator; rotation there is fastest; accordingly, centrifugal force is greatest. As a result, partons that are more or less beyond 60^0 latitudinal, north or south, tend to lose their offsetting ability and are pulled in toward the proton.

The proton's surface is the edge of our 3-sphere world. It encloses the hypersphere and, in so doing, allows for both its stability and very existence. Hence, to all normal objects "within" the universe, the "other" side to the proton's surface is "outside" the universe. It is, essentially, the fourth-dimensional abyss. Partons attracted and captured disappear into that imaginary space. However, some, by just missing the proton, are flung out into a north–south trajectory. They resemble the comets of our solar system. They, then, orbit in a vertical, north–south direction, while the overall electron cloud circulates in a horizontal, east–west rotation.

The vertical orbit delineates our magnetic lines of force. When broken, an electric current forms. Thus, a simple electric generator consists of a coil of wire rotating through a magnetic field. Partons pulled into a proton's interior enter the higher dimension, then circle back as magnetic lines of force. The energy they carry is from this higher realm—zero point energy. We consider it electricity and have learned how to tame it for useful work.

The geometry of the hydrogen atom is stable. Since all of

—

the more than one-hundred other elements are merely combinations of hydrogen atoms, with nuclei consisting of both protons and neutrons (hydrogen atoms from opposite sides), all matter assumes the same overall configuration—central nuclei with vast electron clouds. The larger the nucleus, the more complex the shape; but all are essentially the same.

THE IMPREGNABLE VORTEX

Partons of space that strike a rotating electron cloud, due either to gravitational attraction or inertia, do not disrupt that configuration. They merely encroach upon it, pushing it in the opposite direction. All objects on our planet's surface are constantly being bombarded; the impinging force of accelerating space is gravity. Let me draw this:

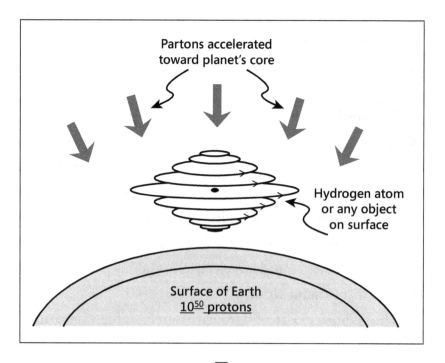

Partons accelerated toward planet's core

Hydrogen atom or any object on surface

Surface of Earth
10^{50} protons

Likewise, if an object is accelerating, either on a planet's surface or in outer space, that acceleration is a force that is resisted by a counterforce—Newton's third law. As a result, partons counter the accelerating electron cloud. Thus we get the following:

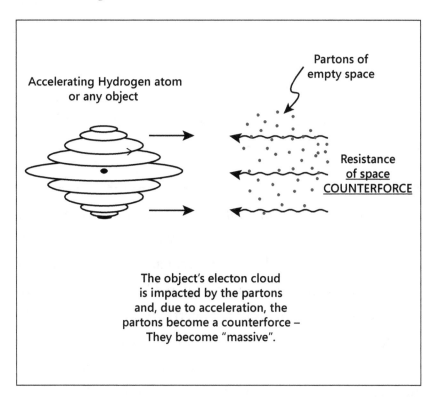

Accelerating Hydrogen atom or any object

Partons of empty space

Resistance of space COUNTERFORCE

The object's electon cloud is impacted by the partons and, due to acceleration, the partons become a counterforce – They become "massive".

UFOs

A vehicle that uses antigravity essentially assumes the configuration of a hydrogen atom. Thus, UFOs take the common shape of a disc or saucer. This is similar to an electron cloud; if drawn we get the following:

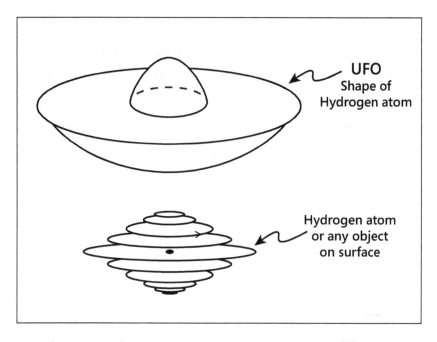

The UFO, then, puts out a surrounding electron shield. That "armor" prevents incoming partons from impacting the space within the vehicle. As the vehicle accelerates, zipping about, it is therefore no longer hindered by any counterforce. Its occupants are not subject to inertia or gravity. Thus, the UFO, by using a powerful *capacitor* (a device that separates charge), in essence, takes that flying disc out of *our* three-dimensional space. It now exists in its *own* three-dimensional space, unencumbered by the battering of partons. (Please see the illustration on the next page.)

UFOs use means other than capacitors as well to separate charge. By rapidly spinning plasma (matter, so highly energized, that electrons are detached from their protons), loose electrons can be dispersed to guard the exterior of the craft.

Also, by using a powerful magnetic field, equal in strength to an electron shield, the space vehicle can be insulated from the bombardment of partons.

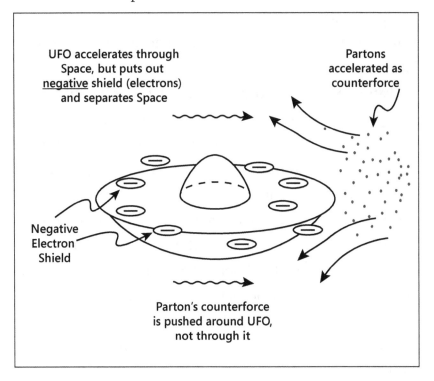

So there are several ways to accomplish the same end. It is believed that the United States, Russia, probably China, and possibly other advanced countries have achieved antigravity flight. Although the power sources employed to propel these vehicles are unknown, it is thought that some means of drawing energy from the vacuum—zero point energy—is in use.

Ben Rich, a past director of Lockheed's "skunk works," the division of today's Lockheed Martin Corporation that is associated with their most advanced projects—the U-2 surveillance aircraft, the radar evading F-117 stealth fighter, and the SR-71 Blackbird reconnaissance aircraft, to name a few—said in the 1990s that we already had the ability to "take ET home." Thus, over 25 years ago we were flying, faster than light, to the stars. One can only imagine how much more we have advanced.

Patrick Shanahan, former President Trump's initial choice to replace Jim Mattis as Secretary of Defense (domestic violence allegations prevented his actual assignment), wrote an op-ed for the *Wall Street Journal* in May 2019 explaining the necessity for establishing an American Space Force. It was begun in December 2019 and now, as part of the air force, is the sixth branch of the military. (It is generally thought that, through the venue of the Space Force, the government will begin to open its "unconventional" research to public scrutiny.) The interesting thing about Patrick Shanahan is that he had been recruited from the aerospace industry and his previous position had been as a leader of Boeing's "advanced projects" division. Boeing, like Lockheed, makes very unusual craft besides their commercial and military jets; those exotic vehicles most likely employ antigravity technology.

Chapter 18

ZERO POINT ENERGY (ZPE)

ZERO POINT ENERGY IS THE ENERGY HIDDEN in the vacuum of space. Space is particulate. Each parton of space is the smallest yet the most massive entity possible. Empty space teems with energy. Unusual particles constantly form, then almost instantaneously disappear. Waves of electromagnetic energy course through this arena. They mostly cancel one another, as they come from all directions.

However, Hendrik Casimir, a Dutch physicist in the late 1940s, theorized that these waves of energy would cause certain observable effects. He predicted that, if two flat pieces of metal were put close enough to each other without touching, these waves of energy would subsequently force them together.

He was advancing a concept in physics that was similar to what is seen in ships at sea. When two large vessels approach close enough side to side in the open ocean, the normal action of waves causes a collision. They are unable to avoid contact. The ocean waves on the seaward sides of both vessels push them together, while the absence of waves between them (blocked by the boats) cannot keep them apart. Let me draw this:

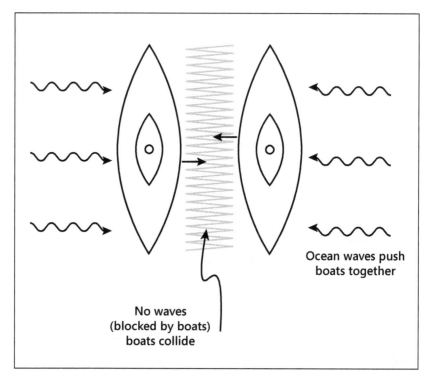

Ocean waves push boats together

No waves (blocked by boats) boats collide

Casimir felt that the vacuum, with its constant, random waves of energy, would act the same and be noticeable if similarly blocked. The experiment was later carried out and found to be as predicted. Therefore, the energy of the vacuum, frequently alluded to as zero point energy, exists. Furthermore, it is immense—10^{120} times what is observable.

We have already discussed this with the generation of electricity. There, magnetic lines of force broken by a rotating coil of metal reform as electric currents. These magnetic lines extend north and south from the atom's nucleus. They are due to the proton's attraction overcoming a weakened centrifugal counterforce, pulling partons into the fourth-dimensional interior of the proton. They then "circle" back, returning as pure energy (ZPE).

We are already taming ZPE as electricity. (It probably is also being used in more advanced governmental projects, but these are closely kept secrets.)

If the vacuum is teeming with waves of energy and vanishing, ephemeral particles, one can begin to explain the unusual findings in Einstein's special theory of relativity. He showed that, when velocity increases, length contracts, mass increases, and time expands. The reason is due to the granular nature of the vacuum.

FORCE OF ACCELERATION
AND THE DOPPLER EFFECT

As an object accelerates, a counterforce is established. Partons, suddenly "awakened" by the acceleration (force) of the transiting object, are energized (become massive) and push back (inertia). Their mass hits the circling vortices of energy (electron clouds) that surround all protons, squeezing them front to back. We get, therefore, a contraction of the length of the moving object. The more rapid its acceleration, the greater the force and, hence, the stronger the counterforce—Newton's third law.

Therefore, when something is traveling at a rapid pace, that object accelerated at some point to reach its current speed. As a result, it was squeezed or foreshortened. It cannot still occupy the same length as when at rest. In a similar manner, its mass increased; more atoms now fit into the same volume of space. The object is denser than when not traveling.

Once it attains its rapid velocity and is no longer accelerating, it also does not return to its original length. Empty space is not quiescent; it is really a sea of energy waves scattered in all directions. These waves cancel one another and, therefore, are not readily noticeable when the object travels at a slow, constant pace. However, if the object moves at a rapid, relativistic velocity (close to the speed of light), the waves of energy striking it head-on are significantly foreshortened (compared to those hitting its back). This is a version of the Doppler effect, similar to the high pitch of an ambulance's siren when approaching, and its lower pitch when departing.

The shorter the wave, the higher is its frequency or energy. Hence an imbalance occurs, with a higher energy density at the front and a lower density at the rear. For that reason, atoms moving forward continue to have their electron clouds squeezed into tighter vortices. Let me draw this:

—

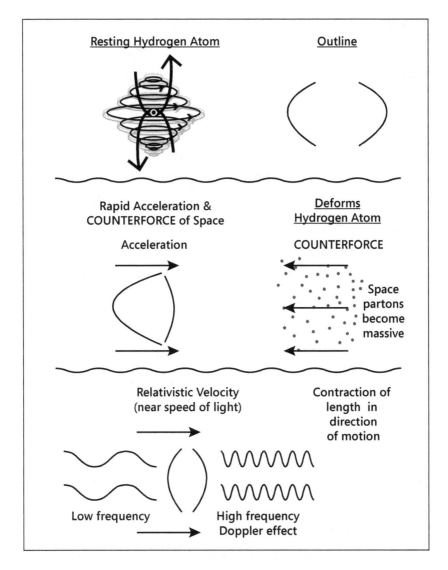

VELOCITY OF REGENERATION—DISTANCE/TIME

The object's time similarly expands. Time is how we measure the constant reestablishment of the universe every Planck moment (5.4×10^{-44} second). There are about 20 million, trillion, trillion, trillion such occurrences every second,

—

far too many to count. However, the universe must regenerate in each Planck moment, as it only exists during that moment. Although time appears continuous, it, just like space, is particulate.

Velocity is measured as *distance/time*—for example, miles/hour or meters/second. Each particle has a fixed size. The universe is simply the largest "particle." All must regenerate each Planck moment; otherwise the world does not exist. Therefore, the greater the size of the particle, the faster is its speed of regeneration.

The entire universe, at 10^{26} meters, consequently reforms at the greatest velocity—10^{26} m/5.4×10^{-44} s, or about 10^{69} meters/second. The electron cloud encircling a hydrogen atom, at 10^{-10} meters, reestablishes at a much slower speed—10^{-10} m/5.4×10^{-44} s, or more-or-less 10^{33} m/s. A proton, at 10^{-15} meters, reforms at a slower speed yet—10^{-15} m/5.4×10^{-44} s, or around 10^{28} m/s. Finally, the smallest possible "particle," the parton, reforms at the slowest possible velocity—1.6×10^{-35} m/5.4×10^{-44} s, or 3×10^{8} m/s—the speed of light.

Although all velocities are different, they all are, essentially, instantaneous. They are all based on the same instant. The speed of gravity is the fastest in the universe (instantaneous regeneration of the largest particle—the universe); the speed of light is the slowest (instantaneous regeneration of the smallest particle—the parton). They differ because of unequal distances covered.

TIME DILATION

We can measure time by the use of any standard, recurring interval. It can be the constant ticking of a clock (seconds), the rotation of the Earth on its axis (days), the orbit of the Earth about the Sun (years), or the time for light to travel a Planck length (Planck moment). Therefore, when observing the time dilation of any moving object, one first notes that its length has contracted. Hence, its velocity of reformulation (its size divided by a Planck moment) has also decreased. As a result, there is a longer interval for reestablishment; that object would take longer to come and go a given distance.

What has taken place is that its speed, into and out from the fourth dimension, has slowed. The "distance" traveled has not. (Although the length of the moving object has shortened, the distance it has to travel—through the imaginary fourth dimension—is always constant.) Thus there is a longer tick-tock of the clock, a longer time for the coming and going of the object in motion. Its time has expanded.

Einstein's *special* theory of relativity, therefore, can be understood by the impact of partons on accelerating atoms. The distortions in length, mass, and time are all caused by moving through the sea of "empty" space. In a likewise manner, a fundamental aspect of Einstein's *general* theory of relativity can also be explained. Einstein theorized that time slows down in a gravitational field—the greater the gravitational force, the slower the time. On Earth, for example, a person living in a high, mountainous area ages at a slightly faster rate

—

than one living at sea level. The difference, although minute, is real.

The reason is exactly analogous to what was just shown. The closer to the surface of our planet, the greater the gravitational force; hence, the greater the acceleration of space particles (partons). As a result, their "awakened" mass when striking an electron cloud causes a more significant contraction of length. This in turn leads to a slower velocity of reformulation, or a slowdown in the internal tick-tock of that object's clock. Time has slowed.

Therefore, Einstein's theories are Newton's basic laws applied to a granular world.

INSTANTANEITY—GRAVITY AND THE SPEED OF LIGHT

The special theory of relativity is based on the constancy of the speed of light. The speed of light is bound up with the concept of a granular universe. Zeno's paradox demands particulate space if motion is real. It was shown by the physicist Max Planck that there is a smallest size to things— since known as the Planck scale.

Along with granularity in size, there is also granularity in time. Things exist but disappear and reappear more than 10^{43} times per second. Each "particle" that makes up our world must, therefore, reformulate about 20 million, trillion, trillion, trillion times every second.

If a wave of energy is moving through what is called

"empty space," it interacts with partons that then interact with their neighbors, ad infinitum. However, the speed of the continual "bumping" is constrained by the appearance and disappearance of the three dimensions of our world. So, since an electromagnetic wave is a pulse of energy through partons of empty space, it is measurable, rapid, and constant. It is the speed of light—the instantaneous speed to regenerate the smallest "particle."

Gravity, however, is the reestablishment of the entire world each Planck moment. Just as the parton is the smallest possible particle, the universe as a whole is the largest—it is everything. The same instant compels a regeneration of each particle, from the tiniest to the grandest. Hence, both the speed of light and the velocity of gravity are the same. They both demand the reestablishment of reality in each moment of existence. Gravity, therefore, is a field of energy that "always" exists. It is the residua of every proton. It is what is found beyond the electron cloud, and it extends to the edge of the universe.

However, in the unusual circumstance where there is a *rearrangement* of existent mass—in the collision of black holes or neutron stars, where two become one—the disturbance caused is transmitted just as is an electromagnetic wave. These occurrences have recently been observed with LIGO (Laser Interferometer Gravitational-wave Observatory). Hence, in the occasional loss of mass as seen in these encounters, a wave of gravitational energy, similar to an exceedingly weak electromagnetic wave, is transmitted through the partons of space at the speed of light.

However, it is much weaker than the electromagnetic wave. An electromagnetic wave is caused by the loss of the protective electron cover. The full force of the proton is then felt beyond that cover, pulling on partons of space. These in turn pull on their neighbors, and the wave propagates outward. But really it is a pull *in* toward the proton. It is comparable to a wave on a still pond after a stone is dropped. The force is a pull in toward the closing hole formed by the stone as it sinks to the bottom. Thus a pull toward a center leads to an expanding wave.

A gravitational wave is much weaker than an electromagnetic one, as the electron cloud has not been disrupted. The full force of the proton's attraction is not felt. However, the jostling of partons is similar and, therefore, like an electromagnetic wave, it advances at the speed of light.

Chapter 19

REASSESS

"*Some* type of phenomenon that is the result of *something* that we don't yet understand and could involve *some* type of activity that *some* might say constitutes a different form of life" *(emphasis added)*.
—*John Brennan, former CIA director, concerning unexplained "UFO" sightings. Jennifer Jett, "Why Are We All Talking About U.F.O.s Right Now?" New York Times, June 3, 2021.*

E MPTY SPACE IS *NOT* "NOTHING." It is made up of the universe's smallest "particles." Because of this, it interferes with any object's movement. It is the cause of the distortions in length, mass, and time found in Einstein's theories. Furthermore, when the motion of space is accelerated, it is sensed as either gravity or inertia. Therefore, if we follow Newton's basic laws of physics (force equals acceleration, and force demands

—

counterforce) and add particulate space, we discover what Newton and Einstein had tried to come to grips with.

We know that Newton had misgivings about the concept of any force acting without direct contact. We see that concern in his letter to Bishop Bentley. We also know that Einstein had a similar foreboding concerning his own theories. This is acknowledged in the letter to his close friend. Einstein's worrisome comments most likely arose from his intimate knowledge of the Roswell UFO incident. He saw that his concepts had fundamental flaws. The world simply did not act as he had thought. Beings from other stars had actually visited us. They, therefore, could travel interstellar distances and would have, most likely, journeyed faster than light.

Finally, Ben Rich, the director of Lockheed's famous "skunk works" would remark to visiting fellow engineers that we already had the capacity to "take ET home." He would famously note that they had discovered a "flaw in the formulas," referring to Einstein's equations. Tesla's allusions to Einstein's theories, therefore, were spot-on. They *were* the "purple robes" that hid a "beggar" masquerading as a "king." Although Einstein's math was valid, his concepts proved illusory.

Where do we now stand? Why does the United States Space Force actually exist? What and where does it fly? Whom, or what, does it protect us from?

IMPORTANCE OF ROSWELL

The Roswell, New Mexico, incident was real.* It also was a great deal more significant than most imagine. President Truman put James Forrestal, his newly appointed secretary of defense, in charge of delving into and capitalizing on what was found and, equally as critical, in keeping it secret. The great fears were that, if made public, the Soviets would learn of and try to co-opt our very considerable find; and, almost as importantly, acknowledgement might lead to a profound panic as the *War of the Worlds* broadcast, by Orson Welles, had just nine years earlier.

Therefore, Secretary Forrestal set up Majestic-12, a group of twelve high-ranking military and scientific leaders entrusted both in maximizing the knowledge and, at the same time, keeping it secret. This led directly to the formation of the Central Intelligence Agency and the National Security Agency. Thus, for the last seventy-plus years, multiple layers of secrecy have been erected in order to conceal the existence of extraterrestrials and UFOs.

The Roswell UFO craft was far more advanced than anything we were capable of, or even considering, at that time. It allowed for major scientific breakthroughs that today we take for granted. In an intriguing book by Phillip Corso, *The Day After Roswell*, he explains how the United States military, in partnership with both industry and major teaching institutions, back-engineered lasers, infrared vision, transistors, integrated

---

*I urge you to examine Appendices B and C, starting on page 115, obtained via a Freedom of Information Act request and other means. They are, to say the least, illuminating.

—

95

circuits, and fiber optics. All of these major scientific accomplishments and more, including the high-strength fabric Kevlar, the binary code for computers, and the magnetic levitation of trains (maglev), were taken from that and later, similar craft.

GREATER IMPORTANCE OF AZTEC

Although Roswell was thus very significant, more important yet was another incident that followed. In 1948, a space craft was found entirely intact, sitting on its tripods, in an area outside of Aztec, New Mexico. It was entered (by finally breaking through its outer surface); inside were initially found two dead extraterrestrials, believed to be its pilots.

An inner section was then opened containing a total of fourteen "caskets," each cooled by liquid nitrogen. Most housed infants (about six months to one year of age), but several had adults, and two were empty (most likely for the dead pilots). Realizing that the others were still alive, the scientists attempted to awaken them by initially experimenting on the infants, feeling that, if unsuccessful, they could still try to revive the adults. They finally succeeded, and three infants and one adult survived. The adult appeared similar to you and me and, most interestingly, spoke excellent English. Over the next year, while being kept on a military base in Los Alamos, New Mexico, he gave us an extraordinary amount of information. He, then, was transferred to a "safe-house" in Vermont and, about six months later, returned "home," when his compatriots arrived to take him back.

He explained that others of his "kind" had been visiting

Earth for many thousands of years. They had helped previous civilizations with building advanced structures, some of which were still extant. He further said that they had come this time because of "signals" that had arrived at their star sent inadvertently by Nikolai Tesla, while he was doing his extraordinary work in Colorado in 1899. They had answered Tesla, but we, at that time, were not knowledgeable enough to interpret their response. Therefore, they felt it was necessary to further investigate the cause.*

Secretary Forrestal felt that this invaluable information

---

*According to the account given by the extraterrestrial, they had "immediately" answered Tesla's signals with similar "energy waves." Unlike light, this energy traveled "instantaneously." This information is corroborated in the book Tesla, the Life and Times of an Electric Messiah, by Nigel Cawthorne (Chartwell Books, 2014, New York, NY, p. 86). It contains a thorough discussion of Tesla's Colorado Springs endeavors. Tesla had apparently connected a telephone receiver to his apparatus that beeped each time a signal was detected. He was astonished to hear regular beeps—first one, then two, then three. He felt that they could not be purely accidental but, instead, had to be under intelligent control. Although he could not decipher them he stated that "the feeling was constantly growing in me that I have been the first to hear the greeting of one planet to another. . .[he later exclaimed] 'Brethren, we have a message from another world, unknown and remote. It reads: one. . .two. . .three.'" The extraterrestrial further stated that, when Tesla did not respond to their signal, they decided to travel to our vicinity—a journey that took about 20 years. (He said that the trip was not unusual for them, as they had planned to come to this area anyway, and that a trip of such length was not bothersome to individuals who live as long as they do—thousands of years.) In 1922 (upon arrival) they contacted Marconi (the inventor of radio) using the letter "V" (his own code for visitors). Marconi told everyone that a signal had come from outer space; however, nobody really listened. They then concluded that Tesla had been way ahead of the rest of our science—even by accident, he should not have been able to figure out how to do what he did in his lifetime.

—

should be made public. The rest of the Majestic-12 group, however, disagreed. As a result, they had him confined to the Bethesda Naval Hospital on the basis of an alleged mental breakdown and, several months later, he supposedly committed suicide by jumping from an upper-story window. However, a special investigative panel later concluded that, most likely, he had been tricked, drugged, or pushed to his death.

MAJESTIC-12

Majestic-12's treatment of their founder and leader, a highly prominent government official, showed to what extent they would go to keep the extraterrestrial information secret. In general, similar actions have been followed for the last seventy years. If someone claims to have other-worldly information that person is considered crazy. He or she is ridiculed as believing in "little green men." If, however, that individual has real proof and plans to go public, he or she disappears.

President Eisenhower, in the mid 1950s, felt that the Majestic-12 team was getting out of control. He understood that Area 51, in Nevada, was where secret military vehicles were being built and tested. He also knew that, some miles away, in area S-4, the more advanced work of back-engineering UFOs was ongoing. He discussed this with Allen Dulles, the younger brother of his secretary of state (John Foster Dulles), who headed the CIA and Majestic-12, and demanded to be brought up to date. Dulles supposedly told Eisenhower that he did not "need to know."

This so enraged Eisenhower that he threatened to have the First Army, stationed in Colorado, "invade" Area 51. Dulles then backed down and gave Eisenhower complete knowledge of what was being done. Because of this and similar incidents, Eisenhower warned America, in his farewell 1961 speech, to be wary of the "military-industrial complex." The speech underlined the seriousness of the problem that loss of control of this very advanced technology could bring.

KENNEDY ASSASSINATION

When John Kennedy became president, Eisenhower fully briefed him on these problems. Kennedy, however, did not have the clout that Eisenhower, a five-star general and the former Supreme Commander of Allied Forces in Europe, had. Kennedy had nevertheless been closely associated with James Forrestal, who, besides being the secretary of defense, had also served as secretary of the navy during the latter half of World War II. Kennedy had been in Naval Intelligence then, and had served as an important aide to Forrestal. He, therefore, was aware of what had happened to his prior mentor and wanted very much to get to the bottom of the UFO problem.

Allen Dulles, however, still in charge of the CIA, was unwilling to give up control of Area 51 or S-4. Kennedy subsequently had Dulles removed from his leadership position at the CIA; however, real power still rested in Majestic-12. In order to make an "end-run" around the CIA and Majestic-12 and take control of our space technology, Kennedy then entered into a pact with Nikita Khrushchev, the Soviet leader, concerning joint space operations. This was finalized in November 1963 and would have forced Majestic-12 to cede control to NASA and the military. Two weeks later, as is well known, during his visit to Dallas, Kennedy was assassinated. Although the Warren Commission claimed that there was only a single gunman, Lee Harvey Oswald, many Americans have their doubts.

REAGAN AND STAR WARS

Ronald Reagan became president in 1981 and, during his first year in office, was attacked by an "obsessed" gunman and seriously wounded. Some in the UFO community think that this may have been a "warning" not to delve too deeply into the CIA and its control of alien technology. Near the end of Reagan's second term, in a speech to the United Nations, he noted that although there had been disagreements and strife with our cold war adversaries, if we—the people of the world—were to be threatened by a hostile, other-worldly force, we would most certainly band together to defend our planet.

The speech was in reference to his "Star Wars" deployment, which, though allegedly designed as a defense against Soviet and Chinese missiles, was, in fact, also intended to counter any potential attack from unfriendly extraterrestrials. Supposedly, our space forces were quite sophisticated by the 1980s, and our missile shield capabilities included advanced particle beam weaponry. Reagan was furthermore reported to have stated midway through his presidency that the United States could deploy upward of three hundred astronauts, at any time, in orbit. Since we were then, allegedly, only flying the space shuttle, a vehicle that could take at most eight to ten individuals at a time, there obviously had to be other venues, both for transport and accommodation.

DONALD RUMSFELD AND 9/11

Donald Rumsfeld was George W. Bush's first secretary of defense. He noted in a September 10, 2001, speech that the Department of Defense could not account for a shortfall of $2.3 trillion. The next day, as we all know, the 9/11 attack occurred.

This immense, unaccounted for budget deficit, almost eight times the "known" spending for the entire Department of Defense (approximately $300 billion a year at that time), was never again seriously looked into. There are those who think that the entire 9/11 attack, with about three thousand lives lost, was an attempt to shift the focus from Rumsfeld's concerns.

Rumsfeld was also attempting, at that time, to set up a military space force. He wanted to wrest the control of space back from the CIA and Majestic-12. This was put aside following 9/11 as the United States went to war against militants and terrorists in both Asia and the Middle East.

When Donald Trump became president, he brought back

Rumsfeld's initial consideration of a Space Force. Finally, in December 2019, the sixth branch of the United States military was formed.

The United States stopped the Space Shuttle program in 2011. For the last decade we have been using Russian rockets to ferry astronauts into space. We stopped going to the Moon in late 1972. Is it conceivable that the United States, the strongest military in the world, has no ability to fly in space? Is it possible that only Elon Musk can successfully fly and land rockets?

WHAT WE FLY

The United States Navy released videos of their F18 pilots chasing UFOs. (In Navy parlance they are "UAPs"—unidentified aerial phenomena.) Many think that the UFOs in these videos are actually flown by the air force. The space fleet must have some means of getting to space. Since it is currently a branch of the air force, perhaps they are using these or similar UFOs.

However, the real reason that the navy was willing to allow this highly classified information out was that, in its upper echelons, it was well known that the service had even more advanced antigravity vehicles. In fact, it is believed that the navy has fleets of kilometer-long space vehicles, similar to their seagoing aircraft carriers, stationed in deep space. It is alleged that these highly advanced, secret vehicles will be shown to the air force once the Space Force becomes an entirely separate division of the military. At that time the navy will fully join with the air force and release its classified weaponry to an independent Space Force.

What then is really out there? What, or whom, are we defending ourselves against?

EXTRATERRESTRIAL BIOLOGICAL ENTITIES (EBEs)

From conversations with the Aztec EBE (extraterrestrial biological entity) and other encounters over the last seventy years, we are aware of basically four types of EBEs.

The most important are the Earth-like humanoids. They appear in most respects to be the same as us. In general, they have no hostile intentions. They look upon our world as we would look upon a wildlife preserve. They are interested in what we are doing but will not interfere with our actions or protect us from ourselves.

They have, however, informed us that there are other forces in outer space that are "hostile" in the sense that they would exploit us although not conquer us. The humanoid EBEs' presence on Earth is to prevent us from doing any "damage" to *their* cultures until we have morally evolved as a species. That is why there was so much UFO activity in the late 1940s, when we first started testing nuclear weapons. Most are hopeful that, over the next several generations, we will become advanced and civilized enough to take part in their galactic space-faring "federation."

The next EBE type is the small humanoid, or "Grays"— so named for their skin appearance. They are essentially the workers or slaves of the other humanoids. They have been genetically engineered to perform the day-to-day tasks required by their more advanced masters. They function like a team of horses would in pulling a wagon. They do not reproduce. In a sense, they are biological robots.

Other than these two humanoid groups, there are those that evolved in non-human or non-mammalian form. We have had little interaction with these groups. Although they "generally" wish us no harm, being entirely different genetically (reptilian or insectoid), they do not consider human welfare or life important.

Lastly, there is a "transmorphic" form of entity. They are entirely different from anything in material form. They exist on another "plane" or dimension and, to us, would appear as pure energy. Supposedly there has been only one encounter with these "beings," brokered for us by an Earth-like humanoid (similar to the one from the Aztec incident). These transmorphic entities can take on any material shape or appearance. They would appear, essentially, God-like.

VOYAGE TO THE STARS

The Earth-like humanoids live and travel in very large craft. These are powered by a form of energy that Tesla discovered when he inadvertently sent signals to nearby stars. These ships are thrust from, or pulled toward, the stars they wish to reach. When approaching a solar system, they, due to their vast size, are "parked" in a distant orbit. Much smaller craft (UFOs) are then dispatched, similar to the ones that were found at Roswell and Aztec, New Mexico. These, in turn, are powered by a form of electrogravitic propulsion in which a gravity wave is balanced against an electromagnetic wave, forming a third kind of "wave" similar to that which powers their larger craft. The occupants, during this part of their journey, are kept in suspended animation. Although they are in a sleep-like state, they still function mentally and can interact with their pilots (EBE grays), if needed.

These voyagers have left their home planets and now traverse the galaxy essentially in search of knowledge and new experiences. In many ways their travels are like those of the *Starship Enterprise* in the television series *Star Trek*. In fact, the producer, Gene Roddenberry, supposedly had access to this information, and his shows are believed to have been a form of "soft disclosure" to the public.

SPACE TREATY

Finally, the United States has officially recognized the extraterrestrials. They are not considered a threat of any kind. On July 18, 1954, President Eisenhower, at Kirtland Air Force Base, New Mexico, signed a treaty with a joint committee of extraterrestrials. The treaty stipulates that the extraterrestrials will refrain from making their presence known, will not abduct or keep any individual against his or her will, and will share with us technical or scientific information in a manner useful to both sides. We, in turn, assigned the extraterrestrials an embassy compound (similar to what was depicted in the movie *Men In Black*) at Majestic-12's facilities in Nevada. This treaty has been acknowledged and maintained by each succeeding President.

CONCLUSION

THE WORLD IS MUCH DIFFERENT THAN YOU OR I have been led to believe. We are but one small planet in a vast universe. Our civilization is just at the start of a space-faring future. We have been guided by more advanced "humanoids" for thousands, perhaps millions, of years. Much of our current technology is based on what was found and back-engineered from these highly developed groups. Einstein was made aware of this information. He was shown the Roswell flying saucer. His self-doubts, as noted in the letter to his close friend, were due to the obvious reality of this advanced technology and faster than light space travel.

Today, this is all just beginning to become public. The possibilities are enormous. New energy sources to end the pollution of our planet will soon be available. Our children, and their offspring, will be living in a far different and better world.

Appendix A: Dark Energy

As already noted, all surfaces, all spheres, are circles ($2\pi r$) multiplied by interiors two dimensions less than what is covered. Hence, the universe as a 3-sphere ($2\pi^2 r^3$) is really just a circle, and a tangent to it can be reduced to our simple example of a one-dimensional straight line.

Thus, if we take the visible universe of 13.8 billion light-years to be a quarter of a circle (0^0-90^0) and divide it equally, we get constant distances two-dimensionally that continually elongate one-dimensionally. Let us, therefore, divide ¼ of a circle into 5 equal portions; each is 18^0 ($90^0/5$) and equals 2.76 billion light-years (13.8/5). However, although these portions are always the same, the tangents increase as we go farther out. The following illustrates this concept:

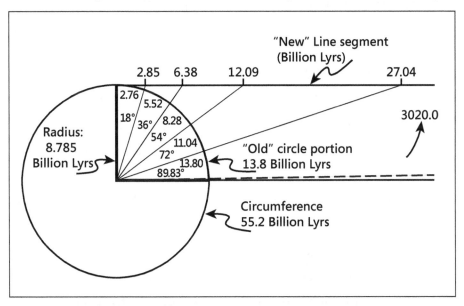

A *tangent* is defined as "opposite/adjacent in a right triangle." In our example, "opposite" is the new length on the straight line, and "adjacent" is the radius of the circle comprising our universe. To find the radius, all we need is the total circumference of our circle. Since 13.8 billion light-years is ¼, the total is 13.8 x 4, or 55.2 billion light-years. Now, $2\pi r$ = circumference, or, if we cross-multiply, r = circumference/2π; since π = 3.14159, 2π = 6.28318; therefore, radius, or r = circumference / 6.28318; or, r = 55.2/6.28318; or, r = 8.785 billion light-years.

Thus, because tangent = opposite/adjacent, tangent = straight line/radius, or (cross-multiply):

Tangent x radius = straight line.

Hence, all we do is multiply the tangent of each angle by 8.785 billion light-years.

ANGLE	TANGENT	RADIUS		LINE SEGMENT	CIRCLE PORTION
18⁰	0.32492 x	8.875	=	2.85	2.76
36⁰	0.72654 x	8.875	=	6.38	5.52
54⁰	1.37368 x	8.875	=	12.09	8.28
72⁰	3.07768 x	8.875	=	27.04	11.04
89.83⁰	343.774 x	8.875	=	3020	almost 13.80

We now have the straight-line segments, and to get the z values we use the formula:

Z = new – old / old

Where "new" is the line segment, and "old" is the circle portion (each 2.76 billion light-years). We now have a z for every tangent, thus a z that matches each distance for the 4D curve/tangent column.

The distances of the Hubble/expansion column are based on the following formula:

Z = square root (1 + v/c / 1 − v/c) − 1.

To get the velocity (v) using Hubble's constant of expansion (the current best estimate is 67.8 km/sec for every 3.26 million light-years [megaparsec]), we take a distance (let us say 1.5 billion light-years) and divide it by 3.26 million to get the number of times its velocity has increased, then multiply by 67.8 k/s for each increase. Thus, in the example, 1.5 billion light-years is approximately 460 (1.5 billion/3.26 million) times 67.8 k/s, or about 30,000 k/s. We then put this velocity into the above formula:

Z = sq rt (1 + 30,000/300,000 / 1 − 30,000/300/000) -1; or, z = sq rt (1.1/0.9) − 1; or, z = sq rt (1.22) − 1; or, z = 1.1 − 1; or, z = 0.1.

Thus, for any distance we can get a z value according to Hubble's law of increasing velocity with distance. When we compare the two columns, we find distance to be greater for the 4D curve/tangent than for Hubble/expansion at every z until about 2.0.

Hence the universe is really bigger, because of a fourth-dimensional curve, than thought due to simple expansion. Dark energy, therefore, has no rationale. It is not required, as there has been *no initial or additional expansion.* It is an illusion.

Appendix B:
The White Hot Report

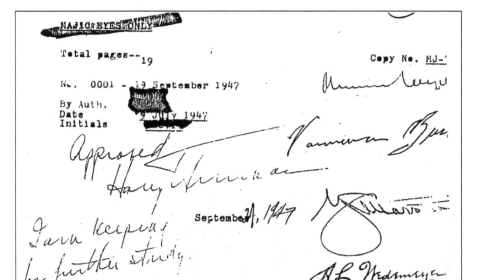

NOTE: No one, without express permission from the President, may disseminate the information contained in this Report or communicate it to any unauthorized person not possessing MAJI SECURITY CLEARANCE.

Those authorized to disseminate such information must employ only the most secure means, must take every precaution to avoid compromising sources, and must limit dissemination to the minimum number of secure and responsible persons who have a "need to know" in order to discharge their responsibity

No action is to be taken on information herein reported, regardless of temporary advantage, if such action might have the effect of revealing the existence of such information foreign intelligence agencies who might exploit for reasons of endangering national security interests.

Foreign powers not amicable to to strategic national security interests will attempt to exploit such information in the possession of the United States Government. They do not know and must not be permitted to learn, either the degree of our accomplishment or the particular source from which any technological advances ████████████████████████████ in this report.

MAJIC EYES ONLY

SUBJECT: MISSION ASSESSMENT OF RECOVERED LENTICULAR AERODYNE
OBJECT AND IMPLICATIONS IN ... PARTS.

... .

PAR. ... TECHNICAL EVALUATION

PAR. ...—SCIENTIFIC PROBABIL. ...

PART IV.—POLITICAL CONSIDERAT...

V/E . NATIONAL SECURITY STRUCTURE

MEMBERS OF MISSION

> Captain Charles ... Buchanan
> Colonel A. Robert Ginsburg
> Captain Richard P. Glass
> Carlisle H. Humelsine
> Major General Stephen ... Chamberlin
> Major General George ... McDonald
> Rear Admiral Paul F. Lee
> Major General Leslie R. Groves
> Brigadier General E.K. Wright
> James S. Lay, Jr.
> Thomas J. Lynch
> Lieutenant General Natnan F. Twining

CONTRIBUTING MEMBERS:

> Dr. Theodore von Karman
> Dr. J. Robert Oppenheimer
> Dr. James H. Doolittle
> Professor Albert Einstein

MAJIC EYES ONLY

4

SIF 7/27/86   0020122

19 SEP 47
24

MAJIC ~~EYES~~ ~~ONLY~~

GENERAL STATEMENT

went out over   try's
1. On 24 June 1947, there ~~went~~ ~~out~~ ~~over~~ ~~this~~ coun'
newswires, with simultaneous broadcast by nationwide radio, a
                                                           of
most sensational story. A civilian pilot by the name of
          while flying his
Kenneth Arnold, a businessman of ~~~~ ~~~~ ~~~~, a salesman
~~fire-fighting~~ equipment, ~~while~~ ~~flying~~ ~~his~~ personal plane on
~~~~ ~~~~ ~~~~ ~~~~ ~~~~ ~~~~ ~~~~ ~~~~ ~~~~ ~~~~ ~~~~ ~~~~ ~~~~
~~~~ ~~~~ ~~~~ ~~~~

~~~~ ~~~~ ~~~~ ~~~~ ~~~~ ~~~~ ~~~~ ~~~~ ~~~~
9200 feet over a mountain range ~~~~ ~~~~ ~~~~ ~~~~ ~~~~
~~~~ seen nine disc-shaped, tailess, silvery-colored object
aircraft, flipping and flashing through clear sky. The
aircraft, if they were such flew like a "saucer" would
~~~~ skipped ~~ across the water ~~~~ ~~~~ ~~~~ ~~~~
white flashes from their surfaces. They flew in formation
~~~~ ~~~~ ~~~~ ~~~~ ~~~~ ~~~~ ~~~~ ~~~~ ~~~~ ~~~~

"Cascade Mountains" their speed had been estimated to be
in excess of 1,000 MPH. The pilot was questioned at length
by Army Air Force intelligence officers and FBI agent. So
far, he has stuck to his story and refuses to comment further
regarding his sighting. Analysis of his report conducted by
the Intelligence Department of the Army Air Forces Air Material
Command, suggests that some kind of aircraft may have prompted
the report as given by the pilot.

3. Another sighting similar to this one was reported

~~MAJIC~~ ~~EYES~~ ONLY

ℓ

45

0020123

the same day by a Portland prospector by the name of Fred
JOHNSON
Johnson. According to interrogation report Mr. Johnson claims
to have observed six to nine disc-shaped objects flying rapidly
over the Cascade Mountains. He watched the objects through a
small telescope which he used in his prospecting. The report
states that he noticed the compass hand osciated wildly from
side to side, indicating some kind of magnetic disturbance was
.. associated with the objects.

4. On 2? June, according to interrogation of of AAF
.. .. conducted th ....formation of discs flew past his
aircraft over Lake Meand ... .. ███████████████████████████
███████████████ the alle ....ers that what he saw was .. ...
way conventional aircraft.

5. Numerous sightings over military installations in
the state of New Mexico were investigated by Army and Navy
intelligence officers. Reports were forwarded to Military
.. ..lligence Division for follow up. ....ties produces ...
..dence that Army and Navy aeronautical research program.. ..
responsible since all such test flights are carefully coordinated
with government agencies and civilian contractors.

6. Up until 4 July, sightings have been reported, not
only in the United States, b.. ... . Germany,
Holland, Paraguay, Scandinavia, Greece, and by ships at sea.

7. Of the 1,200 sighting reports collected since 1942,
approximately 200 incidents have proven to be unknown craft
operating at speeds in excess of 1200 MPH and at times attaining

6

—

120

0020124

altitudes u... w miles or more above the earth. A comparison
was made with Swedish Defence officials of the reported
operating character..... .... & sea
intelligence liaison officia... ... . continuing vie .. ..
of restricted air space over sensitive Swedis: military base
has not indicated any direct .... no tangible
evidence that far .... would sugges ... ... are the
any ....
....

4

0020125

PART I PROJECT WHITE HOT INTELLIGENCE ESTIMATE (PRELIMINARY)

LANDING ZONE NO. 1

Socorro, New Mexico--The unidentified lenticular-shaped aerodyne
which has been designated ULAT-1, has been evaluated as a non-
air breathing aircraft of unknown origin. Totally lacking
conventional wing, fuselage, nacelle, control, and fuel systems
strongly indicates it is not Russian. Consultation with Paperclip
specialists concurr. Aerodynamic features exhibited in ULAT-1
represents a very high degree of engineering and sophistication
not seen in this country. Dimensional homogeneity study cannot
explain how this craft sustains lead and lift factors necessary
for flight. The power plant does not even remotely resemble
any conventional type now in use. Lacking any discernible intake
or exhaust features, it is the opinion of AMC and ONR that this
craft was designed to operate outside of the earth's atmosphere.
The unconventional conclusions reached by members of this fact
finding mission remain tentative at this time. Some members
expressed the view that ULAT-1 may be the product of an advanced
culture from another planet that is much older than ours and has
utilized the science and intellect for interplanetary space
travel. It is not precisely known if the occupants purposely
had the objective of exploration out of curiousity, or with the
intent of surveying for other reasons. So far, no hostile
action or intent has been observed since they made their

8

0020126

presence known. Given the fact that our atomic bomb tests, atmospheric exploration with rockets, and ████████ ████████ ed in New Mexico, could have precipitated the events that led to the incident and subsequent actions taken by the military.

Operating under the assumption that the fallen object was a long-range Russian reconnaissance platform collecting aerial photographic intelligence data, military intelligence personnel were instructed to secure the craft, debris and its occupants as rapidly as possible. Concerns over possible exposure to civilians of unknown biological and chemical agents dictated the quarantine measures taken. Radiation hazards were assumed and appropriate protective measures were taken as well.

In the interest of National Security priorities it was necessary to detain civilian witnesses for interrogation to satisfy intelligence requiements, and quash rumors that could alert potential espionage agents known to be in the vicinity.

Several bodies were discovered. Because on-site medical personnel were unsure of the physiological and biological make up of the occupants, special preparations and preservation methods were employed. Autopsy information obtained so far suggests that the occupants mimic the featuers associated with ORIENTALS, ████████ Outwardly, they appear human-like with but one exception, autopsy notes mention a rarely observed ████████ ████████ s present which supports the premise that these beings originate from another planet.

PART II TECHNICAL EVALUATION (PRELIMINARY)

ULAT-1

    1. Upon close examination of the exterior surfaces of
the craft's fuselage, metallurgists found the skin to be of a
ferrous metal white in color. The metal exhibits all the
characteristics of high-grade steel. It was determined that
the steel was cold-formed and heat-treated. Tensile stength
was estimated in excess of 150,000 pounds per square inch.
Shear tests give the metal a durability rating above 175,000
pounds per square inch, making this fuselage extremely strong
and heat resistant.

    2. Static and pressure flow simulations were impressive.
The low profile ratio of 6-to-1 gives the aerodyne a great
advantage in overcoming the restrictions of the boundary
layer effect in high performance operations.

    3. Spar flanges are constructed in unusual kinematic
design which is believed to allow strain relief at supersonic
speeds. There were no visible signs of plate-stiffeners.
There were no fasteners, weld, rivets, or fittings, holding
the fuselage together.

    4. Lack of wings, flaps, stabilizers, and surface
control features, suggests that the craft is a lifting body.

    5. There are no air intakes or exhaust.

    6. There are no cables

124

0020128

7. There are no identifiable electronics (wiring, ignition, lights, instrument, compartment, engine, motors, vacuum tubes, solenoids, generators, heaters, etc.)

8. The power plant (severely damaged) ██████████ ████████ neutronic engine. ████████████████ detected. heavy water and deuterium (light hydrogen) elements appears be the primary ignitor. A series of coils and heavy magnets connected to the neutronic engine via an oddly arranged group of electrodes (metal not yet identified) appears to be the motive force. One small motor was examined. It is encased in a pure aluminum capsule directly underneath the main eng compartment. There is a small exhaust aperture attached that has what can be only described as an helicoid mechanism ████ ███████████████ The auxiliary motor may be articulated.

9. Navigation and engine contols may be activated tactle manipulation. Viewing may have been achieved by form of television imagery. Symbolic notation appears to be the form of flight and control indicators. Flat panels of unknown metal has been suggested as a device associated with the operation of the aerodyne was discovered and analyzed. It's mode of operation and purpose is unknown.

The absence of provisions, berthing compartments and storage areas, suggest the notion that this craft may be a short range reconnaissance platform. The only recognizable

5

MAJIC

features examined were five loca... ...ents with new machi... fashioned for vertical and be... ...ight. A leaf ... ...beak of a 2'bel... ...nature ... ...nt... ...feat... ...recept...

MAJIC EYES

0020130

Mode of operation is believed to be instrumentation and suggests that the aerodyne from reconstruction of available wreckage ███████████████████████ biosensory and optical stimuli for these reasons:

    a. Absence of indicator lights;

    b. Absence of any circular dials;

    c. Absence of linear dials, or moving pointers;

        Absence of counters;

    e. Absence of scopes;

    f. No mechanical signal indicators

12. There were no identifiable control types found among the assortment of artifacts that would indicate the operation of the propulsion unit was manually activated. knobs, push-buttons, toggle switches, levers, balls, handwheels, handcranks, or foot-pedals were observed in interior space of the flight cabin.

13. The apparent lack of additional clothing and equipment reinforces the belief that the occupants were engaged in a purely exploratory flight.

14. It is not presently known if electromagnetic radiation effects from the power plant had contributed to pilot error or death before impact. If inadequate shielding was the primary cause of pilot error, ████████████████████ s detected.

15. It is believed by some of the crash inspection personnel that sudden decompression and change in atmospheric pressure may have contributed to pilot error. Clothing removed from occupants do not resemble any pressure suits currently

53

Peter Strassberg, M.D.

0020131

being tested by the Army or the Navy for extreme altitude
experiments. Since temperature and humidity factors for the
occupants are unknown, it is impossible to determine if
decompression and temperature change affected circulatory
and dexterity functions. Although it is believed the occupants
may have been overcomed by some yet to be discovered pollutant
or noxious fumes originating inside the craft.

notation or possibly rapid oscillation could have
been a contributing factor in pilot error. It is not known
if organic effects played a part either since since medical
data is non-existant in which to make any judgement as to the
exact cause of death or machine failure.

17. The most probable cause put forward to date for
the crash is believed to be excessive acceleration combined
with steep descent. The seating arrangement was transversely
designed about the vertical axis of the occupants in a positive
direction of flight. The panels removed from the craft
resemble the ones taken from the occupants, suggesting a
symbionic relationship between operator and the functions of
the aerodyne's operation. A very tentative working theory
was expressed by the scientific members of the inspection team
that pilot-aerodyne interaction may occur via electronic-
nonword symbols percieved through the tactile manipulation
of the fingers, feeding impulses to the brain and visa versa.
All of which may suggest a non-inert quality of the materials
existent as being a product of artificial intelligence.

128

0020132

18. The following elements were analyzed and found to exist in the small neutronic power plant that was found inside ULAT-1:

    a. UF6 in metallic form;

    b. hydrogen-fluoride gas;

    c. water and uranium tetrafluoride;

    d. powdered magnesium and potassium chlorate;

    e. metal similar to lead with a chocolate brown color;

    f. U-235 in metallic form;

    g. plastic-like material similar to NE 102;

    h. Beryllium;

    i. pure aluminium;

    h. Thorium isotope material;

    j. Plutonium powder.

19. Scientists from Los Alamos and Sandia Base were alarmed that the power plant could possibly function as a bomb if the elements described above were processed in similar fashion as was done for the lens and shotgun detonators. This originally was the first conclusion. After further evaluation, it was determined that since no recognizable firing circuits were identified, the threat of detonation did not exist.

20. The only evidence of circuitry found on the motor was thin plastic-like sheets fashioned like platters emossed on the exterior of the spherically-shaped casing coated by a thin film of pure silver. Under high power magnification it was observed a series of fine grid-like lines intersecting groups of dots arranged in circular patterns.

5

129

0020133

PART III   SCIENTIFIC PROBABILITIES

    1.   Based on all available evidence collected from
recovered exhibits currently under study by AMC, AFSWP, NEPA,
AEC, ONR, NACA, JRDB, RAND, USAAF SAG, and MIT, are deemed
extraterrestrial in nature. This conclusion was reached
as a result of comparisons of artifacts ████████████
discovery in 1941. The technology is outside the scope of
US science, even that of German rocket and aircraft development.

    2.   Interplanetary space travel is possible provided
adequate funding, necessary resources are made available,
and national interest is piqued.

    3.   Our solar system is not unique. Chances are
favorable for intelligent life on other planets notwithstanding
similar development not unlike our own.

    4.   Being that our culture is relatively young (in
relation to the cosmic scale), it is possible that other
cultures may have developed faster, or are much older and
have avoided the pitfalls common in our historical and
scientific development.

    5.   Human origins may not be constrained to one
planet. Our genus may be found among solar systems similar
to our own.

    6.   The laws of physics and genetics may have a
genesis in a higher, structured order than once previously
thought.

6

PART IV POLITICAL CONSIDERATIONS

1. Given the existing political climate in the US and the unstable conditions in Europe, it is the considered opinion of the members, that if the Administration went public with the information as found in this report now, the results would be damaging, even fatal to the world political structure as it now exists. The following considerations were reviewed and debated which led the mission to the following opinions:

   a. Public trust of the political institutions may be eroded and possibly be held in disrepute.

   b. A complete revisioning make take place among institutions of higher learning, thus calling into question the certainty of scientific knowldge.

   c. The ability of the Armed Forces to secure National Security would be put in jeopardy and possibly lead to undue public fear and disorder.

   d.

   e. History and religion in the political context would probably suffer the most damage causing unprecedented upheaval in social and psychological well-being.

   f. Political repercussions may occur in our diplomatic efforts of containing the Communist threat to our democratic interests.

   g. If such an announcement were made by the current Administration, it could be percieved by opposing party as a trick, laying open to accusations of unethetical posturing and manipulation of the public's mind.

57

0020135

PART V NATIONAL SECURITY STRUCTURE

    With the passage of the National Security Act of 1947, ████████████████████████████████████ ██████████████████████████████ is presented an unprecedented situation with regard to maintaining secrecy related to the information contained in this report.

    In the early months of 1942, up until the present, intrusions of unidentified aircraft have occaisionally been documented, but there has been no serious investigations by the intelligence arm of the Government. Even the recovery case of 1941 did not create a unified intelligence effort to exploit possible technological gains with the exception of the Manhattan Project. We now have an opportunity to extend our technology beyond the threshold that we have achieved, ████████████████████████████████ ████████ Aside from technological gains, we face an even greater challenge, that of learning the intent of such a presence. There are questions that remain unanswered, such as: What forces face us? What kind of defense do we have? Where do they come from and what kind of weapons do they possess? Where can we stage our forces in advance, ████████████████████████████ How wide a front? How many craft can we expect? And ████████████████████ ██

B

—

132

0020136

The members of the mission are prepared to submit a seperate report on just this problem alone. And it would take a dozen volumes to explain how these problems should be met. Our only point, however, is that a combined intelligence and research operation would be a vast, intricate, covertly planned marshalling of resources, human and material, to solve a specific, clearly defined problem.

We have to find effective methods of persuasion with other government agencies without creating a sense of impending doom. The first task is to carefully appraise the problem. The second is to evaluate the known resources and probable strategy of the visitors. The third is to inventory our own ways and means, ascertain how much resources we can bring to bear, and how fast. The third is to devise our strategic plan. And last is to work out with infinite pains the tactical details and the myriad secondary problems of funding and security.

It is the unaminous opinion of the members that Operation MAJESTIC TWELVE be a fully funded and operational TOP SECRET Research and Development intelligence gathering agency. It is also recommended that a panel of experts be appointed to chair and oversee the functions and operations of said agency. It's members should have appropriate security clearances and full cooperation ████████████████████ ████████████████████████ the National Security Council, the Pentagon, ███████████████ t, ███████████████

MAJIC EYES ONLY

5█

—
133

0020137

██████ Joint Intelligence Committee, Joint Intelligence Objectives Agency, Central Intelligence Agency, Atomic Energy Commission, Joint Research and Development Board, Army Security Agency, and the National Advisory Committee on Aeronautics. ███ are highlighted:

1. Propeller-driven bomber aircraft and jet engines, armed with conventional and atomic bombs.

2. Jet fighter aircraft, including some of super-sonic speed, armed with rockets and guns.

3. Propeller-driven aircraft, valued for their endurance.

4. Guided anti-aircraft missiles, and radar-guided anti-aircraft guns.

5. Short and medium-range guided missiles. Drone aircraft.

6. Atomic charges, in bombs, missiles and torpedoes.

In the arena of nuclear weapons we feel there is a certain advantage to be gained ███████████████████ ███████ It is speculated by some that if reduced size and miniture circuitry were introduced into the proposed hydrogen bomb program, it would give US Strategic Air Forces a great deterance capability over the Russians. Current studies at MIT of micro-electronics taken from ULAT-1 may give us the strategic advantage so desired. It is strongly recommended that funding be allocated in this area.

There is a good chance that the Russians may try to make use of the flying saucer scare by public news media

0

134

0020138

MAJIC ~~EYES ONLY~~

and diplomatic means of a technological breakthrough in
aircraft and missile development. We feel that such a
disclosure would most certainly cause great embarressment to
our elected officials and to the military, not to mention
the panic felt by the citizenry. To counter such a threat,
it is recommended that a counterintelligence program be
drawn up and held in abeyance if at such time the situation
should present itself. It might be suggested that we should
make a preempted use of these objects for the purpose of
psychological warfare once the true nature of these objects
are known and understood.

To further assist and aid all MAJCOM in the US and
overseas, it is recommended that a standard intelligence
reporting system be implemented through standard reporting
channels with technical data forwarding instructions. At
present, there are no specific intelligence guidlines
available to military commanders in dealing with sightings
and material evidence collection. It would be advisable for
the respective Secretaries of the Armed Forces to devise a
security policy of plausible denial, if and when the public
becomes aware of the reality of these objects and the
interest of the military of such incidents.

In conclusion, for reasons of national security and
the public well being, the US must be perceived as being the
top of the heap, and every effort must be made to insure
that there is, and never has been, a threat to the country.
~~MAJIC EYES ONLY~~

61

Appendix C:
Operation Majestic

DEFENSE INTELLIGENCE AGENCY
OFFICE OF COUNTERINTELLIGENCE
WASHINGTON, D.C. 28133

THIS IS A SECURITY COVER SHEET

• CLASSIFIED •

ULTRA
TOP SECRET

NO DISSEMINATION OR DECLASSIFICATION

READ-AND-DESTROY

TOP SECRET

SECURITY CLASSIFICATION (WHEN DATA ENTERED)

REPORT DOCUMENTATION PAGE		
1. REPORT NUMBER 405189	2. ACCESSION NO. T-89 EXEMPT	3. MJ-12-739
4. TITLE AND SUBTITLES Assessment of the Situation/ Statement of Position on Unidentified Flying Objects		Natl Security Briefing MJ-12-739
7. AUTHOR(S) CLASSIFIED		405189
9. PERFORMING ORGANIZATION NAME AND ADDRESS Defense Intelligence Agency Office of Counterintelligence Washington, D.C. 20935		405189
11. CONTROLLING OFFICE NAME AND ADDRESS Operation Majestic/MJ-1 ADDRESS CLASSIFIED		08 JAN 1989
14. MONITORING AGENCY NAME AND ADDRESS OFFICE OF THE PRESIDENT		DECLASSIFICATION SCHEDULE T-89 EXEMPT
16. DISTRIBUTION STATEMENT OF ABSTRACT NONE. NO DISSEMINATION/DISTRIBUTION (EYES ONLY MAJIC-12)		

17. DISTRIBUTION STATEMENT OF REPORT
DO NOT DUPLICATE

18. SUPPLEMENTARY NOTES CLASSIFIED

19. KEY WORDS CLASSIFIED

20. ABSTRACT CLASSIFIED

DD FORM 1473 JAN 31 PREVIOUS EDITIONS OBSOLETE

© UNITED STATES DEPARTMENT OF DEFENSE ®

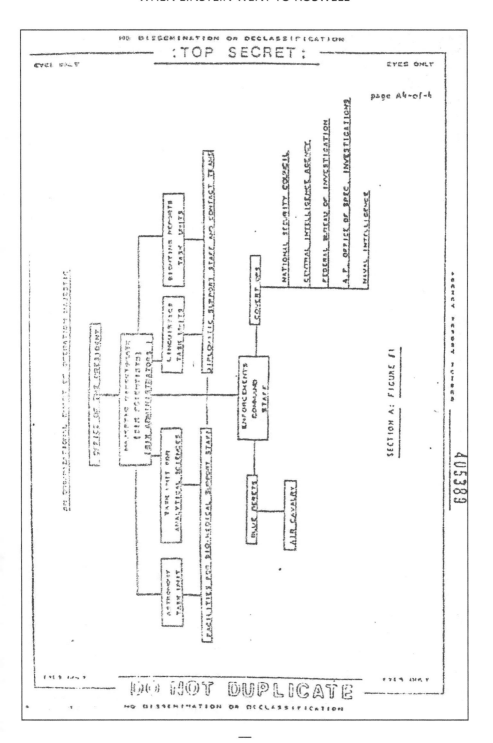

:TOP SECRET:

EYES ONLY   EYES ONLY

HISTORICAL PRECEDENTS

AT

ROSWELL, NEW MEXICO
(07 JULY, 1947)

SECTION B

(6 pages)

EYES ONLY   EYES ONLY

DO NOT DUPLICATE

TOP SECRET

REPORT DOCUMENTATION PAGE CONTINUED	READ INSTRUCTIONS BEFORE COMPLETING FORM

CONTINUE WITH BLOCK NUMBER REFERENCE (FROM OTHER SIDE)

CLASSIFIED

AVAILABILITY-DISTRIBUTION CODES

AVAILABILITY

DISTRIBUTION

SPECIAL CODE

JUSTIFICATION PA11C-12

EXTRA TOP SECRET CLEARANCE REQUIRED REF WHITE HOUSE!!

405389

UNITED STATES DEPARTMENT OF DEFENSE

* TOP SECRET *

NO DISSEMINATION OR DECLASSIFICATION

PRELIMINARY BRIEFINGS

(ENTRY-LEVEL)

FOR

01 JANUARY 1989

ON:

OPERATION

MAJESTIC

CODE NAME:

"PROJECT AQUARIUS"

(COPY ONE OF ONE)

WARNING: This is an ULTRA-TOP SECRET, YOUR EYES ONLY document containing essential compartmentalized information for the national security of these United States of America. Access to the enclosed material is strictly and only limited to persons possessing MAJIC-12 CLEARANCE. Approval is issued by PRESIDENTIAL AUTHORITY ONLY! Reproduction in ANY form or the taking of any sorts of notations is STRICTLY prohibited.

* TOP SECRET *

NO DISSEMINATION OR DECLASSIFICATION

:TOP SECRET:

ⁿ ONLY   EYES ONLY

NOTICE: The ULTRA-TOP SECRET classification utilized in this briefing paper

is a DEFENSE INTELLIGENCE AGENCY/PENTAGON designation. Other agencies and/or

departments of the various military and executive divisions will use another

form of security-level nomenclature. ALL accession is by PRESIDENTIAL

AUTHORITY ONLY, and the following cross-reference of departmental designation

is for the reader's information ONLY.

> DEPT. OF THE ARMY: Top Secret, Red-one (STA/R-1).
> DEPT. OF THE AIR FORCE: Top Secret SIOP/ESI, or, Top Secret SCI.
> DEPT. OF THE NAVY: Top Secret-Ultra, level "S", or, Code Four.
> NATL. SECURITY AGENCY: Top Secret-Ultra, or, Above Top Secret.
> CENTRAL INTELLIGENCE AGENCY: Top Secret, "Red Level-Alpha".
> DEPT. OF STATE: N. S. Provision Exempt/N.S.P.E. Classified.

NOTICE: Within this paper, certain names have been changed or deleted in
order to protect the safety and/or privacy of living persons. This is within
the meaning and scope of the Right To Privacy Act, and/or is being carried
out under the Dept. of Justice's Federal Witness Relocation Program in some
instances. True identities of these persons MAY be provided upon special
request by PRESIDENTIAL APPROVAL and on a NEED-TO-KNOW basis ONLY!

EYES ONLY

DUPLICATE

```
┌─────────────────────────────────────────────────────────┐
│                    :TOP SECRET:                          │
│  EYES ONLY    ─────────────────────────      EYES ONLY   │
│                                                          │
│                                                          │
│              LIST OF CONTENTS BY SECTION                 │
│                                                          │
│   • SECTION A:  Objectives, History and Organization     │
│     of Operation Majestic (four pages).                  │
│                                                          │
│   • SECTION B:  Historical Precedents at Roswell,        │
│     New Mexico (six pages).                              │
│                                                          │
│   • SECTION C:  Historical Precedents at Aztec, New      │
│     Mexico (10 pages).                                   │
│                                                          │
│   • SECTION D:  Selected and Condensed Conversations     │
│     with the Aztec, NM, Extraterrestrial Biological      │
│     Entity (seven pages).                                │
│                                                          │
│   • SECTION E:  Assessment of the Situation and a        │
│     Statement of Position on Unidentified Flying         │
│     Objects.  Prepared for Entry Level Majic-12          │
│     (six pages).                                         │
│                                                          │
│   • APPENDIX :  Selected Illustrations.                  │
│                                                          │
│                                                          │
│  EYES ONLY                                    EYES ONLY  │
│              DO NOT DUPLICATE                             │
│         NO DISSEMINATION OR DECLASSIFICATION             │
└─────────────────────────────────────────────────────────┘
```

:TOP SECRET:

OBJECTIVES

HISTORY

AND

ORGANIZATION

OF

OPERATION MAJESTIC

SECTION A

(4 pages, one illustration)

NO DISSEMINATION OR DECLASSIFICATION

:TOP SECRET:

EYES ONLY   EYES ONLY

page A1-of-4

SUBJECT: OPERATION MAJESTIC-12 PRELIMINARY BRIEFING
PREPARED: 08 JANUARY 1989
BRIEFING
OFFICER: MJ-1

NOTE: This document has been prepared as a preliminary briefing only.
It should be regarded as introductory to a full operations briefing
intended to follow.

OBJECTIVES

OPERATION MAJESTIC-12 (MAJIC-12 or MJ-12) is a ABOVE TOP SECRET
Research and Development/Intelligence operation responsible directly
and only to the President of the United States. Operations of the
project are carried out under control of the MAJIC-12 Group which
was established by special classified executive order # 092447
of President Truman on 24 September 1947, upon recommendation by
Dr. Vannevar Bush and Secretary of Defense James Forrestal.

Objectives of the Group include the evaluating of risk, control
and containment of the violation of United States air space and of
national security by unknown and possibly hostile aliens of one or
more advanced races of beings presumed to have originated elsewhere
than on the planet Earth. The development of effective countermeasures
and the application of any scientific or technical discoveries made
during the operations is also of the highest concern. The following
are some established aspects of the situation:

- Advanced beings of non-human nature are continuously
  being detected along with their flying-disc craft in
  the controlled space of the U.S. since 07 July 1947.

- The remains of seven (7) flying craft and the bodies
  of twenty-seven (27) deceased non-human beings have
  been recovered as of this briefing date, and are at
  present being studied by MAJIC-12 scientists.

- Three of the recovered craft are nearly intact and
  one such machine has maintained some of its power
  since retrieval in 1948.

- Since 1957, no additional alien craft have been
  available for study. The assessment is that these
  beings have adapted or perfected the machines to the
  conditions of our world, so as to avoid any further
  such revealing crashes.

EYES ONLY   EYES ONLY

DO NOT DUPLICATE

NO DISSEMINATION OR DECLASSIFICATION

—

HISTORY

As for the history of the MAJIC-12 Group and its membership and investigations, the complete records are available with proper clearance through the office of counterintelligence of the DEFENSE INTELLIGENCE AGENCY and are on microfilm at this time. The written minutes and transcripts of all directorate meetings as well as a full record of all investigations in their original form are located at the MJ-12MQ, codenamed "Digger Command" and situated at the decommissioned Marine Air Station at Flat Rock, Nevada. This site is now under the covering authority of the Nuclear Regulatory Commission, the story having been circulated widely that this base is involved in the long-term monitoring of residual radiation levels that resulted from the A-bomb testing nearby in the 1950's. With the exception of a monitoring task unit at Wright-Patterson Air Force Base, for the sending up of all sightings relayed through the Dept. of the Air Force, all gathering operations are now located at the Nevada site. Over a period of years, all recovered evidence from the crashed alien craft has been transported to underground storage at Flat Rock, making full use of the large complex of fallout shelters built there starting in 1954.

Scientific and technical research and analysis for MAJIC-12 is being conducted at several places around the country. Small monitoring task units remain in place within the command structure of certain other branches of the armed forces. These include NORAD in Colorado and SAC in Nebraska - for reporting of radar sightings, and SEACOM in Virginia and PACCOM in California for reporting of Naval encounters. The aeronautical research is being carried out by the Redstone Arsenal in Alabama, and linguistic studies are conducted at NASA's Ames research center in California. A team from Johns-Hopkins University and the National Institute of Health is reducing the biological and medical intelligence gathered by the project.

Just as the title implies, MAJIC-12 is headed by a directorate of twelve persons chosen for life (or until retirement), six are military and/or administrative positions, and six are private sector scientists and engineers with long histories of governmental affiliations. All vacancies on the directorate are filled by a unanimous vote of seated directors. The President may recommend an appointment, but while all other decision making lies with this office, membership on the MAJIC-12 directorate is finally decided solely by that group. Such limitation of executive authority is a long-standing historical decision based on the assumption that these twelve persons belong to a small family of legally skilled insiders. Only from within such a family is it possible to have a sense of the "special mindset" required for this very scrutinizing kind of work. It has been said by a former director of the group that the understanding of alien minds requires an alien mind.

NO DISSEMINATION OR DECLASSIFICATION

:TOP SECRET:

EYES ONLY   EYES ONLY

ORGANIZATION

page A3-of-4

The organizational structure of OPERATION MAJESTIC-12 is arranged into three (3) tiers or levels, by order of a need-to-know information flow process. These are:

- **DIRECTORATE** - Composed of twelve members sharing equal authority and complete access to all information. The President is accepted to be the thirteenth member of this body, as well as its final authority*.

- **TASK UNITS** - Composed of a variable number of scientific and information gathering groups and/or persons. The complete listing of current consultants in the sciences is available on request. Identities of Military monitoring units and individuals must be requested through the directorate.

- **ENFORCEMENT** - Composed primarily of selected recruits from the Army Special Forces Command. These units are organized into Companies known as "Blue Berets", and are trained and supported by Air Cavalry units based out of DIGGER COMMAND in Flat Rock, Nevada. Purpose includes containment of tactical situations, control of public awareness, panic, evacuation, etc. Originally created for the security of recovery sights during investigations and the transporting of sensitive evidence. Also includes a number of plainclothed operatives whose purpose is the spread of cover stories to avoid public fears about the inability of the Government to control the situation. These members also operate diversions for civilian investigators to discourage such efforts so as to aid in control of public information and interest.

*NOTE: The identities of the directorate, MJ-12 are held in the strictest secrecy. They are known only to eachother and to the seated President of the United States. These names are never to be written down.

The purpose and need for the highly covert and ABOVE TOP SECRET designation of this operation will become apparent with the study of the following reports which led to the forming of OPERATION MAJESTIC-12. Since the time of these events, over forty years ago, the state of our science has advanced to the degree that we may see a measure of short-sightedness in the then-expedient actions of the administration. As a result of actions taken at that time, several sensitive developments have resulted which are of ever-increasing concern to the National Security of the United States of America.

EYES ONLY   EYES ONLY

DO NOT DUPLICATE

NO DISSEMINATION OR DECLASSIFICATION

:TOP SECRET:

page 0-1-of-6

ROSWELL, NEW MEXICO
(07 JULY, 1947)

On 07 JULY, 1947, a secret operation was begun to assure recovery of the wreckage of an unidentified flying object which had crashed at a remote site in New Mexico. The object was downed approximately seventy-five miles northwest of Roswell Army Air Station (later Walker field), and was reported to authorities there by a local rancher. During the course of this operation, aerial reconnaissance discovered that four (4) small humanlike beings had apparently ejected from the craft at some point before it had exploded. These had fallen to earth about two miles east of the crash site where the wreckage was located. All of the four alien crew members were dead and badly decomposed due to action of predators and exposure to the elements during the approximately one week which had elapsed prior to their recovery. A special scientific team took charge of the removal of the bodies and the recovery of the wreckage of the craft. Civilian and military witnesses in the area were debriefed and many reporters were given the effective cover story that a misguided weather balloon was responsible for the sighting.

A covert analytical effort organized by General Nathan F. Twining and Dr. Vannevar Bush acting on the direct orders of President Harry S. Truman, resulted in a preliminary concensus (19 September, 1947) that the disc-shaped craft was most likely a short range reconnaissance ship. This conclusion was based for the most part on the craft's size and the apparent lack of provisioning. A similar analysis of the four deceased occupants was arranged by Dr. Detlev Bronk. It was the tentative conclusion of this group (30 November, 1947) that although these aliens are generally humanlike in appearance, the biological and evolutionary processes responsible for their development has apparently been quite different from those observed or postulated in homo-sapiens (earth humans). Dr. Bronk's team generated the term, since then widely applied, of "Extra-terrestrial Biological Entities", or EBE's, for these creatures, until such time as a more definitive designation can be agreed upon. It was the unanimous decision of the scientists involved in this investigation that this craft and its entities most certainly did not originate with the government of any nation on the earth.

DO NOT DUPLICATE

Peter Strassberg, M.D.

The crash near Roswell occured sometime in the evening of 2/3 July
1947. At about 20:47 hours at least two dozen persons in the area
observed a bright yellow or "sun colored" disc-shaped object over the
area. On 3 July 1947, in the early afternoon, the widely scattered
wreckage was discovered by local ranch manager William Brazel and his
son and daughter. The authorities at the Roswell Field Army Air Forces
Base were alerted by Mr. Brazel at 09:18 of 07 July, 1947 and two officers
of the base were guided to the crash site by the ranch manager. These
were Major Jesse Marcel, staff intelligence officer of the 509th. Bomb
Group Intelligence Office, and Captain Lee Corns of the Counter-intelligence
Corps.

When the two officers returned to Roswell Field with samples of the
crash site material, they immediately reported to Colonel William Blanchard
of Air Tactical Command. It was at this point that the first of many
decisions were made that have gone into the historical posture of this
government's position on the public's need-to-know about the situation.
Colonel Blanchard released an official press statement confirming that
wreckage of a flying disc had been recovered. This was phoned in to an
Albuquerque, New Mexico radio station without approval from higher level
command in the Army Air Corps. Indeed, no one in the Pentagon or anywhere
else was notified or consulted prior to the release of this information.

In defense of Colonel Blanchard's actions, it should be noted that
no established procedure on the issue existed at the time. There were
no Office of Special Investigations, Central Intelligence Agency, National
Security Agency, or Defense Intelligence Agency or any other similar
such organizations at that time. It should be noted that within less
then two years following this incident, and as its direct result, all
of the above agencies were established with the exception of the Defense
Intelligence Agency which was chartered in 1961.

At nearly the same time on the evening of 07 July, two events were
taking the incident in widely different directions at widely separated
locations. In Roswell, Major Marcel was ordered by Colonel Blanchard
to load the debris onboard a B-29 aircraft and fly it to Wright Field
(now Wright-Patterson AFB) at Dayton, Ohio, for examination. As this
was being done, a reporter from the Albuquerque radio station called to
Wright Field to confirm the crash with the commanding general there.
This was Lt. General Nathan Twining, Air Material Command, where any such
reported incident should have been filed. In fact, the wreckage and all
persons involved were already on their way to Wright Field. But this
was the first that General Twining had heard of the matter. Twining was
forced to state that he knew nothing of a crashed disc craft.

152

:TOP SECRET:

Upon finally receiving verbal confirmation by open telephone line
m Colonel Blanchard at Roswell Field, General Twining did three things
rapid order in an effort to contain the situation. First, he sent
cial orders by secured teletype to Blanchard at Roswell, instructing
to do nothing and talk to no one until further orders from him.
t, Twining used the same system to contact General Roger Ramey, who
in command at Carswell Army Air Forces Base (HQ-8th. Air Force),
t Worth, Texas, ordering him to direct Major Marcel to vector a
ding for an intermediate stop at Carswell, which was on his flight
h. General Ramey was instructed to: A., remove the recovered debris
m Marcel's aircraft and place it onboard another in sealed crates to
forwarded to Wright Field. B., order Marcel and his crew not to talk
reporters. C., notify the press of a photo conference to be held at
'swell where an explaination would be given. D., state at the conference
it the wreckage was only the remains of a weather balloon and its attached
ifoil radar target and prominently display created evidence of same.
, return Major Marcel at once to Roswell Field. Then General Twining
it an open teletype to the newsrooms at the two Albuquerque radio stations
'ning them to cease transmission of the original story and to contact
'swell Office of Public Information for the correct release. Finally,
e general boarded a plane and flew directly to Roswell Field in New
xico.

At his arrival in Roswell, General Twining relieved Colonel Blanchard
command and ordered troops to set up a secured perimeter around the
ient crash site. He then personally supervised the total policing of
e area and removal of all remaining evidence, as well as the four day
briefing of Major Marcel and the six day debriefing of rancher Brazel
m he held incommunicado until the clean-up was complete.

The wreckage recovered at Roswell consisted of three hundred and
enty-seven (327) fragmented pieces of various structural debris. Most
ominent among these were beams of many lengths and ranging in diameter
thickness from just over half an inch, to just under two inches.
ese were found to be composed of a very light and porous metal alloy
mposed of extremely pure aluminum and silicon mixed with zirconium
an unknown crystalline structure. The beams were flexible and were
solutely unaffected by both oxyacetylene and oxyhydrogen blowpipe
tting torches. Several of these contained hieroglyphic and alphabetic
itings of unknown type etched onto their surfaces. Also found were many
ces of a thin metallic foil-like material presumed to have been the
ter casting of the crafts hull, or skin.

:TOP SECRET:

page 8-4-of-6

Following the discoveries near Roswell, New Mexico it became obvious to the intelligence officers working under the newly appointed Secretary of Defense, James V. Forrestal, that a special operations group was going to be needed to coordinate the accumulation & dissemination of data on this subject. This became known as the Majestic-12 Group when President Harry Truman chartered them in September, 1947.

The original directorate of the Group were as follows:

- Dr. Lloyd Berkner - Scientist and Executive Secretary of the Joint Research and Development Board in 1946 (under Dr. Vannevar Bush). Also headed the special committee to establish the Weapons Systems Evaluation Group. Was Scientific Consultant to the Central Intelligence Agency on UFO's during his years with MJ-12.

- Dr. Detlev Bronk - Internationally known physiologist and biophysicist who was chairman of the National Research Council and the Medical Advisory Board of the Atomic Energy Commission (later, Nuclear Regulatory Commission). Also, Chairman of the Scientific Advisory Committee of the Brookhaven National Laboratory.

- Dr. Vannevar Bush - Recognized as one of the United States leading scientists. Organized the National Defense Research Council in 1941 and the Office of Scientific Research and Development in 1943, which led to the wartime establishment of the Manhattan Project to develop the first atomic bomb. After the war, he became head of the Joint Research and Development Board. Was responsible for the recommendation, along with Secretary Forrestal, to President Truman that the Majestic-12 operation was required. Dr. Bush was a principle organizer of the efforts to bring many German scientists, including Albert Einstein, to the U.S. before and after the war.

- James V. Forrestal - Served as Secretary of the Navy before he became Secretary of Defense under President Truman in July, 1947. Among his first acts as the Defense Secretary was the responsibility for organizing the crash recovery of the flying disc from Roswell, New Mexico, which occurred in the same month he took office. He was the first to approach President Truman about the need for an operation like MJ-12, and was in charge, along with Dr. Bush, of setting up an international intelligence gathering organization to monitor world-wide UFO activity. The results of these efforts would grow to include the Central Intelligence Agency and National Security Agency, both of which were founded shortly after the Roswell and Aztec crashes.

:TOP SECRET:

EYES ONLY   EYES ONLY

At a direct result of circumstances surrounding the Aztec, New Mexico recovery, Secretary Forrestal suffered a mental breakdown in March, 1949, & was admitted to Bethesda Naval Hospital under the cover story of needing a "routine physical check-up". While he was at that facility, he is supposed to have committed suicide by jumping from an upper story window in May, 1949. A special committee of counter-intelligence of the Central Intelligence Agency later concluded that the highest probability scenario involves his having been either drugged, tricked, or pushed into his fatal fall. The presumed reason for this involves details of the Aztec crash which will follow in that section.

* Gordon Gray - Assistant Secretary of the Army at the time of MJ-12 initialization and Secretary of the Army in 1949. In 1950, was appointed as Special Assistant on National Security Affairs to President Truman. In 1951, headed the Psychological Strategy Board for Unidentified Flying Object Impact (PSBUFOI), under Central Intelligence Agency Director Walter Bedell Smith.

* Vice Admiral Roscoe Hillenkoetter - Third Director of Central Intelligence (DCI) from 1947 to 1950. Founding Director of the Central Intelligence Agency, established in September, 1947 to direct International information gathering on UFOs. Was covertly placed on the board of directors of the civilian UFO organization, National Investigations Committee on Aerial Phenomena (NICAP) to monitor their activities and control the release of their information to the public. Was forced to take an early retirement after he broke security making the public statement that "UFOs are real and through a process of official secrecy and ridicule, many citizens are led to believe the unknown flying objects are nonsense." In defense of his position, it must be noted that by this time, the events of the Aztec, New Mexico recovery and the death of Secretary Forrestal had depressed and demoralized most of the original members of the investigating group. By 1951-52, MJ-12 had to be entirely reconstructed due to the feelings of many of the directorate that the public had a right to know the facts.

* Dr. Jerome Hunsaker - Aircraft designer and head of the Departments of Mechanical and Aeronautical Engineering at the Massachusetts Institute of Technology. Chairman of the National Advisory Committee for Aeronautics. Was in charge of studies of the machined portions of the debris recovered from the crash sites. Expert in X-ray and microscopic metals analysis.

:TOP SECRET:

- **Dr. Donald Menzel** - Director of Harvard College Observatory and a consultant to the National Security Agency on astronomy. Held a Top Secret Ultra clearance.

- **General Robert Montague** - Base Commander at the Atomic Energy Commission Installation at Sandia Base, Albuquerque, New Mexico from July, 1947 to February, 1951. Was reassigned and transfered following a heated disagreement over National Policy of MJ-12 regarding sightings around nuclear research facilities. Felt the risk to the public health was of greater issue than security and control of the situation.

- **Rear Admiral Sidney Souers** - First Director of Central Intelligence (DCI) from January-June, 1946. Upon initialization of MJ-12, became Executive Secretary of the National Security Council, in order to control domestic security of UFO information. Following the Aztec, New Mexico recovery and related actions, Souers resigned (1950) his positions in national security and his commission in the Navy. He was retained as a consultant on security to the Executive Branch.

- **General Nathan Twining** - The principle "ramrod" for high security in the matter of Unidentified Flying Objects and an opponent of scientific release of data or the involvement of civilian scientists. Twining was an outstanding commander of bombing operations in both the Pacific and European theaters in World War II. In 1945 he was appointed Commanding General of Air Materiel Command, based at Wright Field (Wright-Patterson AFB). Organized Project Sign and Project Grudge (later became Project Bluebook) to control and discredit civilian sightings and civilian investigations. Chief proponent of covert investigation and clandestine contact attempts regarding UFOs and the EBEs (Extra-terrestrial Biological Entities).

- **General Hoyt Vandenberg** - After an outstanding career in the Army Air Forces, was appointed the second Director of Central Intelligence in 1946 until May, 1947. Following the Aztec, New Mexico incident, he joined General Twining in supporting the sensitive nature of releasing any UFO information to the public. When the Air Technical Intelligence Center published its Top Secret "Estimate of the Situation" indicating the belief that UFOs were interplanetary in origin, in August, 1948, General Vandenberg ordered the document burned before it could be distributed to government scientists with the required clearance. This was directly related to the outcome of the Aztec investigation which began in March of that same year.

It is possible to disclose the identities of the original MJ-12 directorate because they are all deceased. It is necessary in order to understand the climate of confusion and anxiety resulting from the second recovery incident at Aztec, New Mexico in 1948.

DO NOT DUPLICATE

NO DISSEMINATION OR DECLASSIFICATION

:TOP SECRET:-

2 ONLY

EYES ONLY

HISTORICAL PRECEDENTS

AT

AZTEC, NEW MEXICO
(25 MARCH 1948)

AGENCY REPORT NUMBER

405389

SECTION C

(10 pages)

DO NOT DUPLICATE

NO DISSEMINATION OR DECLASSIFICATION

EYES ONLY

157

:TOP SECRET:

AZTEC, NEW MEXICO
(25 MARCH 1948)

page C-1-of-10

On 25 March, 1948 at approximately 16:19 (4:19PM) hours, Mountain Standard Time (local time, or LT), a disc-shaped flying machine came down about twelve (12) miles northeast of the small community of Aztec, New Mexico. The controlled landing occurred in a small desert canyon on the private grazing land of a local farmer and rancher.

The approach of the disc was tracked with the aid of three (3) high powered missile tracking radar stations belonging to the recovery network of the White Sands Test Range and located in classified areas of southwest New Mexico. The disc-craft was first observed in violation of the restricted air space of this facility. The disc was flying on a level glide path and on a heading from the southwest-to-northeast at about 3,200 feet altitude when first detected. This track detection resulted in an Air Defense Command Alert that included the scrambling of squadrons of jet interceptors from the nearby Sandia and White Sands bases. Also, per a recent memorandum issued through Operation Majestic, the Commander-in-Chief of Air Defense Command (CICADC) immediately notified Secretary of State General George C. Marshall.

At 17:53 hours LT (5:53PM,MST) the disc-craft appeared to lose both speed and altitude at nearly the exact moment that it was triangulated by returns from all three (3) radar dish antennae in an effort to vector the interceptor squadrons towards their target. At the time, it seemed to be an evasive tactic on the part of the disc-craft. The current assessment is that the microwaves from these long range radars almost certainly evidenced a damage to the power or flight control systems of the disc-craft.

At 18:02 hours LT, the craft dropped below radar some four (4) minutes before the White Sands jet squadron would have reached intercept point (IP) and engaged-to-disc-craft. Estimates later indicated the disc-craft impacted or landed at approximately 18:19 hours LT. The exact position of the landing zone (LZ) was determined within minutes by calculation of the last radar triangulated tracks and confirmed and by a visual fly-over of interceptors from the White Sands base. The LZ-IP was at the bottom of Hart Canyon, a dry wash area about 12.2 miles northeast of Aztec, New Mexico and 2.3 miles northwest of the Cave Creek bend in the Animas River. This was on private land about 1.7 miles north of the state highway and nearly ½ mile west of unpaved county road S-141 south.

At 18:26 hours LT (6:26PM, MST), on 25 March 1948, the crash site was visually confirmed by the commander of the 152 tactical fighter wing and relayed to the Air Defense Command officer at Sandia Base, Albuquerque, New Mexico. This information was conveyed at once to the CICADC who contacted Secretary of State Marshall. By 18:35 hours LT, General Marshall had made contact with the Army Air Force's Interplanetary Phenomenon Unit (IPU), a section of the Army Counterintelligence Directorate. The IPU was maintained and supported out of nearby Camp Hale, Colorado and was the closest group able to secure, recover, transport and store a disabled disc-craft. This action was entirely correct in the interval before a MAJESTIC operation could be activated and the preparations of the IPU group while awaiting a MAJICOPS crew saved valuable time.

DO NOT DUPLICATE

:TOP SECRET:

EYES ONLY   EYES ONLY

page C-2-of-10

The IPU was formed after the Roswell, New Mexico crash recovery and was later absorbed into the enforcement division of OPERATION MAJESTIC at the time that the Department of the Air Force was established as an entity seperate from the U.S. Army.

The disc-craft was recovered and the LZ declared secured by 22:45 hours LT through the actions of an IPU scout team sent to the Hart Canyon area. Meanwhile, General Marshall had contacted Dr. Vannevar Bush - scientific head of MAJICOPS - and a joint IPU/MAJICOPS team was assembled under MJ-12 command for the purpose of investigating and clearing the LZ, or crash site. The local rancher who owned the land and ▮▮▮▮▮▮▮▮▮▮▮▮▮▮ were held incommunicado while the field analysis and subsequent clean-up was conducted at Hart Canyon. Visiting with the rancher at that time and planning to hunt for game animals on his land, were the owner of a local radio repair shop plus a unemployed oil hunter and inventor. These last two parties were later to get together and tell their story to a columnist from Variety Magazine named Frank Scully. This author later published a book in 1950 detailing parts of the Aztec recovery story. This breach of security was occasioned by the use of a cover story for the IPU/MAJICOPS workers that proved to be ineffective. It was decided to disguise the Aztec efforts as an exploration for oil by a fictional drilling company, a fact that was not in agreement with the knowledge of the unemployed oil hunter who knew that there was no oil in the area. Later both this independent oil hunter and the columnist Scully were discredited by facts leaked to a writer at True magazine by covert operatives of the MAJESTIC Enforcement Division. This action was unfortunate but very vital in light of the sensational nature of the Aztec discoveries. Fortunately, Scully labeled this fortune hunting oilman as a "scientist" in his book and during the period of the book's sales, this adventurer was convicted of fraud for trying to peddle a device he claimed would find oil! Mostly by luck, the security breach healed itself and the book was quickly forgotten with only minor help from MJ-12.

That evening of 15 March, General Marshall ordered ADC off of alert status and requested that they send teletypes to the radar stations advising them that there had been a false alarm. This was an important factor here since the disc-craft had landed intact and had the appearance of having been landed under the control of it's operators. No means of entry was immediately obvious and members of the IPU scout team considered it possible that some of the crew of the disc-craft might still be alive inside it.

At 01:20 hours LT, 16 March, the team of MJ-12 scientists assembled by Dr. Bush arrived at Durango Air Field, Colorado about thirty-five (35) miles northwest of Aztec. All members of the teams of IPU and MAJICOPS investigators were then sworn to an Ultra-Top Secret security oath. All personal property and identification was left in lockers at Durango and the teams boarded trucks headed for the Hart Canyon site. It must be remembered that the helicopter was still in experimental development at this stage and trucks were the best means of entering the rugged terrain as well as the fact that they would cause far less attention than the noisy and unfamiliar helicopter, which was, however, used to bring in the initial IPU scout team.

▮ Name withheld to protect a living individual's privacy.

EYES ONLY   EYES ONLY

DO NOT DUPLICATE

NO DISSEMINATION OR DECLASSIFICATION

:TOP SECRET:

EYES ONLY

EYES ONLY

page C-3-of-10

In order to fully follow the course of the Aztec, R.M. recovery and investigation it is important at this point to detail the names and the backgrounds of the members of the IPU/MAJICOPS team before returning to a chronology of the events at Hart Canyon. Only four of these persons were on the MAJICOPS directorate. These were; Dr. Lloyd Berkner, Dr. Detlev Bronk, Dr. Vannevar Bush, and Dr. Jerome Hunsaker. Their biographies may be found in the preceding file of the Roswell, N.M. incident.

The following were selected by the MJ-12 directorate to accompany the IPU/MAJICOPS recovery team to the Hart Canyon landing site:

- Dr. Carl A. Helland, Director and senior scientist of the Geophysics and Magnetic Sciences Laboratory of the Colorado School of Mines. Was instrumental in discovering that sections of the Aztec disc-craft were held together by molecular magnetic bonding along finely polished seams and determined a means for it's disassembly.

- Dr. John von Neumann, former consultant on the Manhattan Project (atomic bomb) in 1943 and a close personal friend of Dr.'s Bush and Oppenheimer from same. Expert in both mathematics and computing machines. His work on the Aztec disc eventually led to the binary computer language used today in all modern electronic computers.

- Dr. Robert J. Oppenheimer, was Director of the Institute for Advanced Studies at Princeton University from 1947 after a distinguished period of service during the war as head of the Los Alamos atomic bomb project. Also, was Chairman of the General Advisory Committee of the Atomic Energy Commission (later Nuclear Regulatory Commission). Acting for his friend, Dr. Bush, who had not yet arrived at Aztec, he selected most of the recovery team members and arranged with the U.S. Navy for storage in their facilities at Los Alamos until permanent sites could be prepared.

- Dr. Johann von Roesler, an expert nitrogen chemist whose work led to development of Nitrogen Mustard Gas, Nitroglycerin derivatives leading to Trinitrotoluene (TNT) and worked with Dr.'s Bush and Oppenheimer on triggering devices for the atomic bomb. Originally included as a consultant on the possible need to gain forced entry to the disc-craft by use of high explosives, he eventually worked with Dr. Detlev Bronk on investigating the liquid nitrogen cooled cryogenic suspension system used aboard the disc-craft by it's crew members.

EYES ONLY

DO NOT DUPLICATE

EYES ONLY

NO DISSEMINATION OR DECLASSIFICATION

page C-4-of-10

• Dr. Merle A. Tuve, worked with Bush and Oppenheimer for
the Office of Scientific Research and Development.
(Manhattan Project) during World War Two. Top pioneer
in radio-wave propagation in the upper atmosphere. Later
was part of the Bell Laboratories team which developed
the principles of laser light physics derived from study
of systems found aboard the Aztec disc-craft.

• Dr. Horace van Valkenberg, was Director of the School of
Metallurgy at the University of Colorado. Expert in
analytical chemistry. Was useful in determining proper
materials handling of components of the Aztec disc-craft
during investigation and disassembly. Later pioneered
science of X-ray crystalography, crystal holography and
electron microscopic metals analysis as well as the holder
of over thirty (30) patents through Arizona State University
in metal alloy processes. Most of the above were the
result of research he began after his analysis of the
structural materials of the disc-craft.

Although not present during the recovery phase of the Aztec operation at
Hart Canyon, it is important we acknowledge the contributions made later
by Lambros Callihanos and William Friedman who are experts in language and
cryptology and whose analysis of the writings and symbols found aboard the
disc-craft led to a degree of understanding of the cryogenic suspension
system used to freeze the ship's occupants. Also, Dr. Paul A. Scherer,
whose supply of refrigeration and nitrogen cooling and pumping equipment
aided greatly in the revival of the suspended occupants aboard the disc-craft
at the field laboratory set up at Los Alamos. Both Callihanos and Friedman
went on to work for OPERATION MAJESTIC for many years in our Covert Operation
Section of the Enforcement Division as deep planted agents within the
National Security Agency. Dr. Scherer led an illustrious career as Research
and Development Director for Air Research Corp.

The IPU/MAJICOPS convoy took routes to the Hart Canyon site that avoided
main roads, and on arrival they set up road blocks and secured a two (2)
mile perimeter around the canyon rim and landing site. In exchange for his
silence, private land was leased from the farmer who owned the field in
his canyon for the construction of a controlled access road to and from
the area. Equipment carrying trucks were camouflaged to look like oil and
gas drilling rigs and personnel involved wore coveralls with the name of
a fictional oil exploration firm created by MAJICOPS for this purpose. Later
a fence was constructed around the perimeter of the canyon and bore the name
of the oil company and warning of blasting on signs attached to it.

When the time came for removal and restoration of the site, trucks were
used that were labeled "High Explosives" and were escorted by Blue Berets
disguised as National Guardsmen. The cover story which was very effective
in keeping the curious away was spread that these trucks were removing large
amounts of nitro-glycerin used in blasting during oil explorations. This
was the reason given for the "National Guard" escort; that the governor of
the state requested the escort to avoid the theft of the explosives.

:TOP SECRET:

page C-S-of-10

Shortly after sunrise on 26 March 1948 the IPU/MAJICOPS team began operations at the Hart Canyon location. Measurements of the radiation levels around the craft and the landing site were taken and proved negative (i.e., nothing out of the ordinary). Some high concentrations of potassium chlorate powder were detected in soil samples surrounding the disc-craft. A fine, yellowish dust composed of particles less than five (5) microns in diameter was detected coating the nearby underbrush. Chemical analysis showed that this was composed of sulfurous substances in an unfamiliar combination having a bivalent, rhombic crystalline structure and was mixed with an unknown and extremely highly electro-positive element in the iodide group. It is now known that this was a form of disinfectant automatically discharged by the craft upon any touch down to a planet's bioactive soil. Linear calibration surveys were taken by theodolite (surveyor's transit) and photographs were taken from every possible angle with a Graflex 4"x5" plate camera equipped with special calibration overlay masks to register dimensions on the films. The ship was determined to be 99.983 feet in diameter and 42.638 feet in height, or thickness, including the 9.451 foot extension of the landing pods supporting it off of the ground. These three round, hemispheric pods beneath the craft had obviously been extended from within the hull during a controlled landing effort. These were measured to form the points of an equilateral triangle 22.3 feet on a side, center-to-center and were 16.45 feet each from the epicenter of the craft's bottom hull. Both upper and lower hull surfaces were smooth with no means of entry or even of propulsion immediately in evidence. The hull was the color of dull, unpolished aluminum and was apparently metallic in nature with a normal surface temperature. No seams, hatches, rivets or portholes were obvious. Photographs taken at closer than fifteen (15) feet were found to be blurred. Later, both infrared and magnetic radiation fields were found to surround the craft and were no doubt responsible for fogging of the films in close photographs. Watches and other ferric (iron-based) tools and instruments left within this zone around the craft were found to become magnetized.

Eventually the team crawled across the upper hull surface with access by a scaffold erected around one side of the craft. Upon conducting a close examination, they discovered a small, hairline crack at the base of a slight bulge in the domed top of the upper hull. Careful study determined this crack to be irregular enough as to be a result of accidental impact with a solid object (much later a civilian witness was debriefed in Albuquerque who testified that he observed the craft strike the side of a cliff along the south bank of the Animas River at about the same time the saucer ship dropped below radar). After banging on the hull and yelling though a loudspeaker failed to receive any response from the presumed occupants of the disc-craft, it was decided to attempt to widen this crack in an attempt to gain entry. At first the discussion was held as to the possibilities of the craft's internal atmosphere being explosive or that our air could be poison for the occupants to breath. Finally, Dr. Bush waved off such debate on the grounds that if some action was not taken immediately, any chance for survival of members of the ship's crew would run out with the passage of time.

:TOP SECRET:

EYES ONLY   EYES ONLY

page C-6-of10

First heat from blowpipes was applied in excess of ten thousand (10,000) degrees and finally diamond-tipped drill bits were used until the crack was at last extended longitudinally and finally widened into a gap great enough to shine a small electric light inside and look around. Another plate of material of similar construction seemed to be solidly blocking the view. However, at a seam along the bottom edge of the panel, another breach was located. It was then assumed that the craft was constructed of both an inner and an outer hull seperated by only a few inches. After many minutes more of effort, the drill broke through into another empty space. Upon extending this opening and looking in, the research team were surprised to discover the interior to be well illuminated with sunlight. It was discovered that the original opening was forced at the lower corner of a sort of large rectangular window or porthole which had been cracked. These portholes were not visible from the outside, but were extended all the way around the compartment in huge curved rectangles and seperated by only a thin structural brace between them. The exception was the cracked one by which the team gained entry; it had turned a milky translucent color like frosted glass. Later testing showed that the entire outer hull was covered in this transparent material that was then blocked off by an opaque inner hull where desired by the ship's designers. This was a thin and inflexible skin barely thicker than four sheets of typing paper and yet direct blows from sledgehammers would not dent or rupture this foil-like substance.

After experimentation with a long, thin rod used to probe at what looked like levers or buttons inside, a panel in the lower hull extended down along what seemed like invisible seams and became a ramp leading up into the craft. When calling out provided no reply, the team obtained flashlights and other equipment and entered the disc-craft. The greatest discoveries were made on the lower level, dubbed the "cargo hold", but the team of investigators did not immediately find the closed door or hatch to this compartment. Instead they found themselves in a small circular chamber about the size of an elevator cab. A spiral ramp of very steep incline led up around a center post to the upper chamber dubbed the "flight deck". Since the team had already observed some details of this upper chamber and had observed the bodies of two (2) small creatures slumped in chairs, it was decided to investigate these first in case they might be alive.

Inside the upper cabin room the team found the bodies of two (2) small humanoids about four feet in height and strapped into seats like those in a jet cockpit (acceleration couches) which faced a row of instrument panels built into a curved countertop circling the cabin just below the portholes they had previously seen from outside. The extraterrestrials were dead. Later study determined that the normal atmosphere inside the cabin was much thinner and colder than our air. When the crack in the windowframe happened the two (2) "pilots" died from sudden increase in pressure and temperature which caused skin hemorrhages like bruises and gave a brownish hue to their skin that first seemed like burns. The seats holding the bodies were found to adjust in size to fit almost any human form and to glide around a track in the floor that followed the instrument panels.

EYES ONLY

—

:TOP SECRET:

page C-7-ol-10

The group soon recognized that the instrument panels were active and some symbols were backlit by a blue-green glow. Some of these were changing as they watched. At this point Dr. Oppenheimer consulted with Dr. von Neumann and suggested that the team proceed with extreme caution since the craft was obviously still under some form of power and it might be activated into operation either by their interference or by electrical automation from within or without by radio control. Dr. Tuve suggested that the team should explore the construction of the controls rather than attempt to operate them on the chance of being able to trace the power to it's source and deactivate same until some understanding of the purpose of each instrument could be determined. This was agreed upon and the work began. Among the first discoveries was that each panel of instruments in the flight deck countertop slide out like a drawer along nearly invisible lines less than 0.2 millimeters wide and in a track or rail composed of a single strip of plastic wrapped in a continuous "S" curve around two metal rollers at the front and rear. There were no ball bearings, springs or motors. And yet each drawer slid both open and shut at the slightest touch of a fingertip as if self-powered. Later it was found that the rear roller bar traveled back-and-forth on a track of it's own as the drawer was operated. This second track was curved so as to balance the force imparted by the bend in the plastic strip and thus created an almost frictionless mechanism. The type of plastic was then unknown to Earthly science, but has since been developed and has what is known as a "molecular memory" of it's ideal unstressed state, to which it will naturally return. Later it was determined that each section of controls or instruments aboard this craft was self-contained as to it's power needs, by some means integrated into it's nuclear structure that gathered electromagnetic and other energy from the space around it; in this case the fields of the Earth. There was no central power system anywhere on the ship and the control, flight and environmental systems drew their power from unknown sources. Instrumentation was eventually found to interconnect and communicate with each other on the molecular level by means of waves or vibrations set up in the atomic structures of each device. No wires or other recognizable electrical components were found. .

The sections of the saucer hull which contained both upper and lower deck cabins were joined along microscopic lines by means of electromagnetic locking action engineered into their molecular structure. When Dr. Bush and a team of military engineers found several interlocking bay-like devices in a cabinet built into the wall, Dr. Helland determined that these could be used to dismantle the craft into sections. At the joints, the operator of the keys, which linked together for different sized tasks, simply passed the tool over the seam. The device, which looked like a tuning fork and had each tine magnetized with opposite polarity, unlocked the joint when passed over it in one direction and rejoined the sections when it was drawn in the other direction. Even after disassembly and prolonged storage, the separate sections of this craft still maintain their power and operating capabilities to this day. All attempts to alter the magnetic fields that connect sections of the craft, performed with the use of powerful electromagnets under exacting laboratory conditions, have failed to change the degree of the interlocking effect. Also, kinetic force (repeated blows) has failed to alter molecular magnetic alignment according to crystal holographic imaging tests recently conducted.

:TOP SECRET:

page C-8-of-10

The entire inner hull structure was built over a light-weight framework of flexible beams supporting the two level cabins and formed into pie-shaped equal thirds around a single, central, vertical shaft or pole running through the thickness of the craft and connecting the upper and lower outer hulls which also divided into equal thirds. The pole running through the saucer was about forty-one (41) feet long and about sixteen (16) inches in diameter. It was a perfect cylinder of a solid carbon/zinc alloy. When maintained or stored in a vertical position, this pole has no special power or ability. But when laid down at the Hart Canyon site on it's side, the pole began to drain electrical current from machines, generators and batteries and to broadcast a powerful ma   which damaged sensitive instruments and magnetized tools. For th   had to be transported in an upright fashion.

Many other discoveries were made, in   ing three different styles of writing; first, a form of symbolic pic   s, second, an alphabetic style of pictograph which seemed to break do   he complex drawings into words, and third, a form that   like the raised symbols in the Braille lettering for the blind but   . Much later, Callhamos and Friedman showed this last form to be a part of mathematical code based on binary counting and read from right-to-left like Hebrew. But the most important discoveries were made on the lower level cabin, w   team labeled the "cargo hold".

While attempting to take apar   els" in the lower entry chamber, the team discovered a hatchway   y that disappeared so well when closed that they over   it earlier.   s was opened and closed by touching a pictograph lo   the wall beside it. This was curved to match the walls of the c   first called the "central chamber" and later, the "airloc   had no frame around it. The mechanism had no motor and worked similarly to the wheel-less magnetic levitation railroads we are currently planning to construc   a technology adapted by study of this door mechanism. Th   entire airlock chamber rotated like a geared cylinder inside anothe   cylinder that formed a center core at the heart or center of the l   cabin, until the two openings in the two cylinders lined up. The material was so thin that the opening in the inner wall nearly went unseen. Inside the cargo hold were found a number of what appeared to be scientific work tables and medical examination tables. Just as Dr. Bronk was about to suggest that the "cargo" routinely carried might be a biological or   human one, Dr. van Valkenberg found a number (14) of small tube-like chambers buil   the wall beside a clam-shell couch that looked like it would close t   tain a human body. This couch was fitted with hoses for filling it with some liquid or gas when the two sections would come   ther and seal. A closer look at the sealed tubes which looked like the doors of clothes dryers at a laundromat revealed that they were a complex form of refrigeration system. Two were empty and twelve contained the bodies of what looked like human adults and small children as well as infants, all frozen as if preserved for specimens.

The decision was made to remove these from what had now been nicknamed "the   ue", and pack them   ice for preservation until a later autopsy cou   determine if they were a   abducted earth humans.

DO NOT DUPLICATE

:TOP SECRET:

EYES ONLY

EYES ONLY

page C-9-of-10

In the process of transfering the frozen EBE bodies from their capsules, in the cargo hold of the disc-craft, to containers of dry ice for shipment, Dr.s Bronk, von Roesler, Scherer and two Army pathologists from the IPU were amazed to discover that one of the small humanoid adults, similar to the "pilots" found deceased on the flight deck, had thawed out with a pulse, limited respiration and had lived in an unconscious state for several minutes before expiring. The shocked scientists then realized that the capsules were actually cryogenic suspension chambers and that the EBE's were still technically alive inside them. Although this concept was almost unheard of in western medicine at that point in time, Dr. Bronk dimly recalled reading about secret experiments conducted in Nazi concentration camps during the war. Such experiments had centered inhumanely on the suspending of the vital life signs and their reanimation by freezing and thawing of the bodies of inmates who were used against their will like laboratory animals.

By this time, the scientists had determined that each section of the ship was able to maintain some operating power even after disassembly. A test showed this to be true for the cryogenic capsules as well. Due to the lack of proper facilities at the landing site, and the lack of a secured holding area for any living EBE's, (or convalescent provisions) it was decided to remove the entire cryogenic section of the craft, and to relocate to better research and medical laboratories on a military compound of some sort before attempting to restore any of the EBE's to consciousness. The story was quickly circulated that, yes, bodies had been found - all deceased, and that these would be removed for later analysis. The bodies of the two small "pilots" were shown to some other members of the recovery team who had heard about them, and then sealed in dry ice chambers, as was the corpse of the EBE who had expired in the cargo hold after thawing. A big display of this was carried out in full view of the other team members, and other dry ice transportation capsules were secretly loaded with dirt and stones before the entire group was loaded into a refrigerator truck labeled "high explosives" for transport. The rest of the craft, in pieces, was loaded onto flatbed trucks, covered with tarpaulins and labeled "explosives". Dr. Oppenheimer arranged for full use and needed remodeling of the restricted Navy Auxiliary Airfield complex at Los Alamos Base. The convoy proceeded at night and by the most unobserved route possible, to these facilities, while units of the IPU/MAJICOPS team assisted by Army Airborne "Rangers" stayed at the Hart Canyon site to affect final clean-up operations. The disc-craft arrived and was put into secured storage for analysis on 5 April 1948.

Two seperate laboratories were set up at the Los Alamos site. One was for the detailed photographing and cataloging of each component of the disc-craft and the effort to decipher the forms of writing found onboard. The second facility was organized around the efforts of teams of doctors led by Dr. Detlev Bronk, to attempt the revival of the craft's occupants. Neither group knew what was the purpose or progress of the other's research. Only Dr. Oppenheimer was in a position, acting as go-between, to see all phases of the work in progress and to disseminate the findings of each to the appropriate investigators within each group. This was done in order to limit the risk of information reaching the public and causing a panic situation based upon a fear of undetectable aliens among them. The thought of riots and murders was on everyone's mind.

EYES ONLY

EYES ONLY

Peter Strassberg, M.D.

Eventually, the medical team was able to resuscitate one adult Earth-like
humanoid male and three (3) Earth-like humanoid infants, all about six (6)
months of age; two male and one female. The rest of the infants and one
more short, grey-skinned, large-headed humanoid EBE, perished in the attempt
to revive them. This was largely the result of Dr.s Bronk and Bush having
decided to experiment first on the infants, who were of little intelligence
value, in the hope of reviving the short EBE and the adult humanoid for a
detailed debriefing.

The adult, Earth-like humanoid male turned out to be, himself, an EBE.
But he spoke perfect English with a slight and untraceable accent and
exhibited many telepathic and psychic skills as well. This EBE was, in his
general appearance, completely human; internally, there were only slight
differences in the formation of the heart valves, pancreas, lungs and he
possessed two livers and an unfamiliar organ where the gall bladder would
be in an earthman. Also, his digestive and gastro-intestinal systems were
simpler and less able to process the wide range of foods that earthmen are
used to. He was surprised to discover the scientists had unfrozen him alive,
but was otherwise undisturbed to find himself in the company of earthmen.
He stated within minutes of regaining consciousness that his only surprise
would have been if the investigators had not chosen to "hold me captive out
of your honest curiosity". After a hasty tele-conference with President
Truman, it was explained to the visitor that if his intentions proved to be
non-hostile and he cooperated in an information exchange, he would be granted
diplomatic status and soon be repatriated to his own kind when the arrangement
could be made. To this he readily agreed, provided he was not asked to give
away any scientific secrets that could alter the course of our natural
cultural development.

All together, the Aztec EBE lived under our protective custody on the
Los Alamos complex for nearly a full year, from late April 1948 - until
March of 1949. After that, he was sequestered at a private safehouse set-up
by Army Intelligence in rural Vermont, during which time he met with the
President and other top government and military administrators, prior to
his being returned to his people in August of 1949. He gave the scientists
and military debriefers a great deal of mostly non-technical information
about his civilization and it's motives for being on our Earth; a total of
six hundred and eighty-three (683) pages of transcripts were made of recorded
conversations. A condensed version of some noteworthy points from these many
debriefing sessions follows at the end of this section.

The EBE saw little harm in allowing us to keep the remains of his space
craft for study, since he felt our understanding of it would only gradually
develop. He did suggest that his people would "probably have to drop one in
similar condition in the laps of the Soviet Union - just for the balance of
things to be maintained.", and, "You are welcome to take this up with any
higher authority you can find willing to listen to you, if you do not approve
of this course". The team of scientists were told that the human-like infants
were destined for our world anyway, and we were welcome to keep them. On
21 August 1949, the Aztec EBE was returned to his own kind at a meeting site
southwest of Kirtland Air Force Base, Texas, and arrangements were made for
a future meeting at the same location, to open diplomatic relations.

NO DISSEMINATION OR DECLASSIFICATION

:TOP SECRET:

EYES ONLY   EYES ONLY

SELECTED AND CONDENSED

CONVERSATIONS WITH THE AZTEC, NEW MEXICO EBE

SECTION B

(7 pages)

EYES ONLY   EYES ONLY

DO NOT DUPLICATE

NO DISSEMINATION OR DECLASSIFICATION

:TOP SECRET:

ES ONLY   EYES ONLY

page 0-1-of-7

CONDENSED CONVERSATIONS WITH THE AZTEC, NEW MEXICO

EXTRATERRESTRIAL BIOLOGICAL ENTITY
(from April, 1948 to March, 1949)

The following conversations were held over an eleven (11) month interval
between various interrogators, labeled here as INT, and the Aztec EBE.
These are reduced from many hours of recorded debriefing, and are included
here to give the reader a degree of insight into the character and the
personality of the subject. These statements are not intended to provide
complete answers to questions. All transcripts of the debriefing and the
original tape recordings of the same are available through special request
(and with proper clearance) from the DIA office of counterintelligence or
the MAJICOPS-HQ, code; "DIGGER COMMAND", re: "PROJECT AQUARI ". These may
be procured through the EXCOMCON-NET liaison office at the Pentagon's Joint
Strategic Objectives Staff Division (National Defense Section).

INT: "What shall we call you, what is your name in your native
         language?"

EBE: "Oh! How even that isn't easy for me because you will not
         get the phonetic pronunciations correct without practice.
         And I am not able to write it down for you because we do not
         use an alphabetic printed language that I could translate
         to English easily."

INT: "Why don't you try it out on us anyway?"

EBE: "Of course; Sethimus (or Setimus, or Seck-tha-mous) will
         do; that is the first of four names I am known by. In
         our written language, this would be shown by a single
         graphic symbol that would explain my entire lineage."

INT: "Why is your written language so different from ours?"

EBE: "Why is yours so different from the Egyptian of thirty
         centuries ago? forgive me, but that is not a simple
         problem. The group I belong to gave up any identification
         with any one planet, or it's culture, before your people
         learned to make fire. There were once thousands of
         languages, both oral and written, in our many civilizations.
         It became important to use symbols to represent whole words
         and even concepts that all of us held in common. Soon your
         culture will do the same with public signboards at first.
         Travel and intercommunications of many peoples will always
         cause this to happen."

INT: "Why are you here?"

EBE: "Specifically to bring you these children. And we like
         trees."

EYES ONLY   EYES ONLY

DO NOT DUPLICATE

NO DISSEMINATION OR DECLASSIFICATION

:TOP SECRET:

INT: "Would you mind expanding on that?"

EBE: "Certainly; in what direction? You see, we would hate to see you blow-up such a pretty little planet; to ruin so many nice trees. If you only knew how hard it is to make a tree from scratch! People are easy - they follow naturally, but not without trees, of course. Green things must be respected above all else except children, and are very much alike."

INT: "What are the intentions of your government by sending you here? WHO sent you and WHY? If this isn't the first time, evidenced by your familiarity with English, why haven't there been official contacts made?"

EBE: "(sigh)In reverse order; What? and be strung-up from one of those beautiful trees? Why? Because you sent us an invitation we could not resist, for our own curiosity, as well as safety. We sent ourselves, or I volunteered - take your pick. And we have no government; we outlawed it, or rather we outgrew it. I suppose you can not outlaw a thing without laws, and those require governments, do they not?"

INT: "I don't understand."

EBE: "I know that. Try laughing. You still will not understand, but you will feel better about your ignorance."

INT: "But that's not good enough! Make me understand!"

EBE: "It will have to suffice. One can not teach until a student wants to learn. How good do you want us to be? I will not harm you. We will not harm you or interfere anymore than you want us to. We are only here to offer what you asked for in a manner we know you will accept; and you did ask us to come and visit."

INT: "When? Who sent for you? Who invited you?"

EBE: "Almost fifty of your years ago, you sent us signals. You were looking for us, and you scared us with a signal you should not have been capable of sending. So we answered. But you didn't understand our answer then, just as you don't now. We do our best to please others. When you didn't respond to our message to you, and you did not come for a visit, we felt that we really ought to stop by and see how you were doing. It has been a long time, and it was the neighborly thing to do, of course. Besides, it was right on the way to another destination; your world, I mean."

INT: "You've been here before?"

EBE: "I know! We were surprised that you had forgotten us!"

DO NOT DUPLICATE

Peter Strassberg, M.D.

INT: "I don't see how we could forget you! Are you serious?"

EBE: "Very seldom. But, yes, _your_ culture, which now dominates this civilization, did forget us. Actually, your people seem to have undergone a period of history in which you were so afraid of old truths, that you erased them in favor of religious fantasy. Some of your people remembered us orally in their legends, such as those who speak the Nahuatl, Narragansett, or Athapaskan dialects. They lived where we last visited. But the high point of their cultures is long passed."

INT: "I've never heard of them."

EBE: "Not surprising, your history shows that the conqueror seldom preserves any history of their victims. I speak of the natives in this land mass; those called Indians; mostly on your eastern and southern boarders and in the ribbon of land that connects yours to the southern continent. Navahos? Aztecs? Incas? The Olmec's and Toltec's cultures? These were all greater than your own civilization at a time when you were burning witches and killing their cats, which brought a plague that killed millions of your ancestors. Of course, you drove them mad and destroyed them with venereal diseases and smallpox; those you didn't slaughter for their gold."

INT: "Oh."

EBE: "And your social pundits speculate and joke about why we don't land in front of your nation's capitol building and shake hands? We are different, not, insane!"

INT: "but we have changed. Western civilization is now the leader in this world; for freedom and humanity."

EBE: "Two things; first, we believe that, or we wouldn't be here now. Also; tell that to the millions of Hebrews your western civilization has destroyed in the past decade, or the millions of Negro families whose sons died to stop the madman Hitler, but who do not have plumbing in their homes so to speak. Yes, you make progress; that is hopeful for you. But emotionally you are very different from us. Not primitive, that label would be pure egotism on our parts, but you are very different and frightening to us. We are trying to learn to overcome those differences before your runaway technology thrusts you upon us without choice."

INT: "We have no space travel capability, how could we harm you?"

EBE: "In our terms, you are very close to that point. Do you have small children? Yes? Well, they will grow up in a world where space travel is taken for granted. And before they make you a grandfather, earthmen will be in space."

171

:TOP SECRET:

INT: "All right, so you have been visiting us for some time. I have no choice but to accept that, even though I'd love some proof. But you still didn't tell me who sent you the message from Earth that brought you back here!"

EBE: "Funny you should connect those two subjects. If you dig in a place I will show to you on any world map, you will find your proof. Unfortunately, I am now in the wrong country for this, but others of my people will instruct the proper government to dig there if you do not mind. You see, in a remote part of the nation you call Yugoslavia, we visited and helped the people there to build a very advanced culture over seven thousand years ago. It will be found. But there is a beautiful mystery in the fact that the person who sent us your first message forty-nine years ago (1899) did come from this nation originally. His transmission was sent from near here, however, in the state you call Colorado."

INT: "Can you tell us his name? Are there any other proofs of your earlier visits?"

EBE: "Thousands of proofs, if you look around. The scientist I spoke of was an electrical researcher named Nikola Tesla. You are old enough to have heard of him when he was still alive. You know, I just thought of a very good puzzle for someone of you to solve: on an island you named the Isle of Pines in your Pacific Ocean, you will find what is left of a concrete landing platform we built there one hundred and fifteen (115) of your centuries ago. You can not miss finding that, since the footings we put down still cover many acres. Look for those if you would like to; no culture on your planet could have built them at that time. It would be fun for me if you ever locate them, because my great-grandfather helped to pour the mortar for them!"

INT: "What! How long do your people live?"

EBE: "Somewhat too long, I think at times. We have built-up our age somewhat. It would vary, but I was alive when your whole continent was practically a question-mark on the sea charts."

INT: "What about these babies we found with you? Are they human? What are we supposed to do with them?"

EBE: "Raise them, of course, as I would have done. Certainly they are human; perhaps more-so depending on whose definition we use. You are too far away from understanding the whole science of bio-engineering yet for this to be sensible, but just accept that we have sewn the seeds of a rich harvest for your culture. And this will continue. They are only children, not monsters. But raise them well, please, because the care of children is one of a few things we do not laugh about."

Peter Strassberg, M.D.

NO DISSEMINATION OR DECLASSIFICATION

:TOP SECRET:

EYES ONLY   EYES ONLY

page D-5-of-7

INT: "I remember this man, Tesla, you spoke of; he died during the war (1943), but I never heard that he tried to contact other worlds."

ESE: "It was in all the papers, along with the story of how he was always being taken to court by his neighbors for creating lightning which struck their properties. Besides, when Tesla wouldn't answer, we sent a reply to your great inventor, Marconi, that we had arrived for a visit."

INT: "Really? When was this? I'm sure he never reported it!"

ESE: "You see how you hear, but don't really listen? Again, it made all the papers. We'd just arrived here in, oh, 1920 or '22 (it was 1922), and we sent him the letter Y in his own code. It meant that he had visitors! He told everyone that he knew for certain that the signal came from outer space, but nobody really listened. Well, after that we began to figure out that your man Tesla was way ahead of the rest of your science! Even by accident, he shouldn't have learned how to do what he did in his lifetime."

INT: "Why did it take twenty years for you to answer Tesla's message?"

ESE: "It did not. We answered it immediately! Oh! I see! Our arrival took twenty-two (22) years, not twenty, because that is how long it takes to get here from where we were when we got Tesla's invitation."

INT: "So long! You took such a far journey just to find us, and didn't even contact us?"

ESE: "As I said, it was on the way to somewhere else we were planning to visit anyway. Besides, how long or far is all relative, is it not, when we live as long as we do? And you do not see even now that we always live in space. We outgrew our origin worlds many generations ago. In fact, you have to try and understand that you are really living in space too. The only difference is that we left our natural planet and built a new home to go where we wanted. You did the same thing when you left your caves and built huts; and for the same reason. So it's all relative."

INT: "But why not contact us after such a journey?"

ESE: (laughing) "I am sorry! But you still do not see! We did contact you. We are contacting you; all to the limits of your ability to hear our message. I keep telling you that it is relative! Please! Laugh about it! I am trying to explain a limit to your perception which is caused by your limited perception! Do you see the wonderful paradox!"

EYES ONLY   EYES ONLY

DO NOT DUPLICATE

NO DISSEMINATION OR DECLASSIFICATION

173

:TOP SECRET:

page 8-6-of-7

INT: "If you say so, but such a trip has no purpose in our way of seeing things."

EBE: "Of course not. That is why we do what we do. Listen now, because this could go on forever if you do not. We are at home where ever we park our craft. It is huge by any standards you could apply. We are contented because this area of space is one we cross regularly as a simple matter of course, like your campers who vacation at the same place every year. Suddenly we find our camping site threatened by bears! But bears are protected, you are not supposed to hurt them; so you must tame them or give up the whole vacation. Do you see that although you are a long way from travel to other stars, by our standards, you will be overrunning the whole area by the next time our children pass by here? Do you see why it is vital, unless you destroy yourselves first, for us to tame you now; before you come to us! Would you let savage headhunters roam the streets of a city without sending them to school first? We cannot interfere with you anymore than you would interfere with bears in your parks. Someday we would rather you joined us, visited our worlds where we do make camp, and joined us on our long voyage of discovery. Or not. It is your destiny, as long as you do not try and alter ours. Try and grow into our perspective, and all the illogic you question will seem very reasonable. It is all a matter of scale. Centuries instead of decades. Someday you will be just about like us, and you will see. Some new, short-lived species will think you are a crazy metaphysical paradox too. And you will have to learn to treat them gently as children, without letting your egos get in the way."

INT: "But with your science, you could make any world your own personal paradise; why live in spaceships!"

EBE: "Yes, and with our weapons we could take what we want as well. But why have one world when you can see them all? You all covet land, a very old idea and only natural when the supply is limited. Just try to accept that we are in paradise when we are seeing and understanding something new. Our god is change and our religion is understanding. Searching for truth is our greatest social challenge and purpose. When you build up your age, this will become as clear to you as why you must breath air."

INT: "It sounds like yours is the ideal, perfect world."

EBE: "No. When you have more years, filling them profitably is a real problem. And the trouble with being a loving species is that you want to prove it by making more of your kind with a partner you love. Of course this must be closely regulated in a culture such as ours, and so relationships of an intimate nature are difficult to leave to one's heart."

:TOP SECRET:

INT: "You speak of the heart being the seat of emotions. This
is a concept we use on Earth, but we know it to be a sort
of romantic fiction. Why does an advanced race like yours
maintain such a notion?"

ESE: "Well, I used that particular expression for your benifit.
But why do you object to romance? Your species is driven
by little else except romantic vision. Why not ours as well?
Certainly our way of life must seem romantic to you? It
does to us. We write poems and express ourselves with
works of art just as you do. Many of the greatest works of
construction on your world were inspired by a joyful effort
of our people and yours, in your distant past. We leave
these wherever we go. And you romantic creatures have tried
for many generations to explain why such huge monuments to
human engineering were ever built in the first place. Why,
they were built to express the joy of living, of course!
And the great sense of satisfaction our people have gotten
from working side-by-side with yours has been one of our
biggest joys in life. That is another reason for our being
here. Surely even a small child can understand such joy?"

INT: "You keep talking about a message for us. Can you define that
a little more?"

ESE: "We simply are. We have no deeper purpose than to be.
We do not wish direct contact with your people. You would
reject us in your present frame of mind. So we seek to be
accepted at the lowest level possible, and hence the broadest
one possible. That is as a cultural myth; we seek to be
the next Santa Claus to put it bluntly. This must come to
your people, from your people. The understanding will be
given you by the widened perceptive skills of these children
we are bringing you. For your sakes this must not come from
political or religious leaders. It must flow evenly throughout
all the common cultures of the Earth. The message is not that
we are here, but that each of you can control where you are
going. Be like us, or different; as you choose. But you
choose, not your political leaders who believe that only a
tiny percent of the elite can rule wisely over their masses;
or those who spread fear of death in order to enrich their
lives through religion. Knowledge is only good when spread
like seeds from which wisdom might grow; and faith has purpose
only when it is not created of fear. With faith in the wisdom
of truth, and in each other, your world will come to ours soon
enough. Our hope is to be able to receive you as friends and
equals when you arrive. That is our message. You must do the
work for yourselves. If you do, you will keep all the rewards."

NO DISSEMINATION OR DECLASSIFICATION

:TOP SECRET:

EYES ONLY

EYES ONLY

ASSESSMENT OF THE SITUATION

STATEMENT OF POSITION

ON

UNIDENTIFIED FLYING OBJECTS

PREPARED FOR ENTRY LEVEL MAJIC-12: (EYES ONLY)

SECTION R

(6 pages)

EYES ONLY

EYES ONLY

DO NOT DUPLICATE

:TOP SECRET:

YCS ONLY   EYES ONLY

page E-1-of-6

ASSESSMENT OF THE SITUATION

In any attempt to properly assess the present situation as it relates to the issue of national security or the risk of cultural upheaval, it is important to take full advantage of hindsight and historical retrospective. It is both an interesting footnote of history and a key point of reference that the human race itself is directly responsible for its being contacted in this century by beings from other worlds.

Through dialogs with representatives of these alien pilots of the flying disc-craft, we have learned that our own scientific curiousity led to the visitations which have been ongoing throughout this century.

In 1899, the Yugoslavic electrical scientist, Nikola Tesla, most noted for his introduction of alternating current to electrical power transmission and for a laboratory device named after him (the Tesla coil) embarked on a series of researches that have made this the saucer century. Tesla had long proposed that it was possible to directly broadcast pure electrical energies at a distance without loss of power and without wires. By 1899, and with the aid of government and private scientific backing, Tesla had chosen a site near Colorado Springs, Colorado to conduct a massive and never repeated experiment. Nikola Tesla's purpose was to gather the Earth's own magnetic field and to use the Earth as a huge transmitter to send signals to outer space in an attempt to contact whoever might be living there. Tesla had no idea that the specific type of power he had generated was coursing through space and caused great havoc many light years away. Modern science has all but forgotten Tesla's work. The sorts of energies and magnetic waves he was working with were never fully defined or explained and almost all of his original designs are lost or were destroyed by their creator in his later years. But it is now obvious that this single action by a scientific outlaw generated serious concerns among the other intelligences in the near reaches of our galaxy. Although others have attempted to use radio waves to broadcast messages into outer space, none used the energies that Tesla did.

We have now determined by discussion with the extraterrestrials who have contacted this government since 1947 that the type of power Tesla used is a very important form of energy. It has apparently been largely overlooked by modern electrical engineers and physicists although certain information gathered by the Central Intelligence Agency (CIA) indicates a strong program of research of this type inside the Soviet Union. According to the alien sources we have debriefed, Tesla's work was a technical fluke at least one hundred years ahead of it's time. Development of these energy fields will only logically occur after we have fully understood laser beam devices.

The extraterrestrials have reportedly been visiting and regularly monitoring the progress of life on Earth for thousands of our years and they were very concerned about this odd signal we sent coursing through space. Among other things, it seems that these energy waves can disturb the process of erology, alter weather patterns and damage ozone levels in whatever atmosphere they might encounter.

EYES ONLY

177

:TOP SECRET:

Evidently, Tesla was sending huge amounts of an unknown form of energy into space along with his intended signals but was unable to detect these due to the fact that instruments for their measurement did not and do not now exist.

The extraterrestrial intelligences (EI's) attempted to respond to his transmissions in a form of binary code that they routinely use for long range communications (evidently these energies act instantly at a distance and are not limited to the speed of light) and ask that he cease sending. Of course, Tesla had no way of understanding the message he received back from space. Fortunately, the anger of local residents at the side effects of his research forced him to shut down the Colorado Springs experiments in the same year he began them.

This, then, was the actual start of the so-called "flying saucer age" in our times. As it became clear that our people were on the verge of an explosion of technical progress, the EI's decided on a long term program of carefully calculated and seemingly random contact with the eventual goal of raising our awareness of our place in the galactic community. With the advent of the Atomic Age, this program was escalated to include eventual diplomatic contact with many of Earth's governments. The same approach of staging apparent "accidental" contact was chosen for it's low psychological impact on the human race. This was the situation in the case of the United States of America (re: Roswell and Aztec, New Mexico files, this briefing).

The problem with direct contact was stated early in our relationship by one of the extraterrestrial biological entities (EBE's), " Right now there are people on this planet who do not know that your moon circles your world or how long it takes your world to orbit your star (sun). I am not speaking of a small number, but perhaps as many as half of the global population. We have been mistaken for being gods or devils in your past and this sort of slow education is best. As a cultural myth, we stand the best chance of acceptance by the masses of your citizens."

There are now hundreds of compiled studies of the EBEs, their worlds and their objectives. In several cases, volunteers from our armed forces have participated in diplomatic and cultural exchanges during the 1970's and early 1980's and have visited some of the EBE's worlds. The following is a short assessment of the situation as we now understand it.

° There are four basic types of EBEs so-far confirmed;
  And they are listed here in descending order of their
  influence on our studies.

  A. Earth-like humanoids. There are several
     variations more-or-less like ourselves. The
     majority of these are friendly and are the
     bulk of our EBE contacts. Most have a high
     degree of psychic ability and all use science
     and engineering of an advanced nature.

DO NOT DUPLICATE

Peter Strassberg, M.D.

NO DISSEMINATION OR DECLASSIFICATION

:TOP SECRET:

EYES ONLY   EYES ONLY

B. Small humanoids or "Greys". The Greys, so-called
for the hue of skin possessed by most of this
type, are a sort of drone. The are not unlike
the worker ants or bees. They are a type of
genetic clone incapable of sexual reproduction.
With little or no independent will, they are
mostly under the psychic control of the Earth-
like humanoids who raise them like pets (or a
kind of slave). A detailed explanation of this
cultural oddity and it's role in the form of
space travel these EBEs employ is addressed
elsewhere in this briefing. Assuming the Greys
are under benign control, they are harmless.

C. Non-humanoid EBEs. These are in several classes
and come from worlds where dominant morphology
took a different evolutionary course. Many of
these are dangerous not for organized hostile
intentions, but because such creatures do not
hold human life as sacred. To them, we are
animals despite our intelligence. A few are
friendly because their cultures forbid all acts
against living creatures. Thus far, contact has
been minimal with only a handful of unfortunate
encounters.

D. Transmorphic Entities. Of all the forms of EBE
studied so-far by Operation Majestic, these are
the most difficult to understand or even to give
a description of. Essentially, such entities are
not "beings", or "creatures"/ TEs exist in some
other dimension or plane which is to say not in
our space or time. They do not use devices to
travel in space. Only once has a member of an
MJ-12 contact team been privileged to be present
when introduced by an Earth-like EBE to a TE. In
essence these entities are composed of pure mind
energy. They are not unlike "ghostly spirits"
and seem to be only curious about our universe.
They are said (by other EBEs) to be capable of
taking on any physical form that they "channel"
their energy into as matter.

* The principle means of travel used by the humanoid EBE is
of a mechanical nature and involves very large craft that
cannot land on a planet. These employ the forms of energy
discovered accidently by Nikola Tesla to cause themselves
to be thrust from or pulled towards the stars they wish
to visit. Some alteration or control of the passage of
time onboard the intransit craft is used as well.

EYES ONLY   EYES ONLY

DO NOT DUPLICATE

405389

179

:TOP SECRET:

- Since the huge interstellar (between stars) craft
  cannot approach the powerful gravity and magnetic
  fields of the stars from which they draw their power,
  they are placed into a "parking orbit" far outside
  the planetary family (solar system) of any star they
  approach. This may be millions of miles from the world
  the EBE's wish to visit. Smaller craft, often but not
  always disc-shaped, are used to travel the distances
  remaining. These draw their power and propel themselves by use of "electrogravity". This is a means
  of balancing a gravity wave against an electromagnetic
  wave in a manner that generates a third kind of wave
  similar to that which powers their larger vessels.
  This journey between worlds takes time in the small
  craft, and the humanoid EBE's suspend the effects of
  aging by freezing themselves in small chambers. The
  psychic powers of the humanoid EBE's are not lessened
  during this state of "suspended animation", so they
  are still able to control their craft. Towards that
  end and to provide for some external control over
  unforeseeable circumstances while frozen, the humanoid
  EBE's developed a host/parasite culture thousands of
  years ago with the small "Grey" EBE's. Since then,
  these Greys have become an extension of their controller's
  senses to such a degree that most humanoid EBE's rely
  heavily on their use at all times. "They are not unlike
  the symbiotic empathy between a pack of hunting dogs or
  a team of horses, and their earthly masters - simply
  like tools or partners in work," explained one EBE.

- During the current wave of interest in our planet
  by the EBE's, there have been three interstellar craft,
  or "mother ships" stationed outside our solar system.
  These have been rotated at twenty-two (22) year intervals
  since the early 1930's. During a contact debriefing
  in 1964, an EBE explained that this was no hardship for
  them. "Our genetic science is about where yours will be
  in another millenia or two. Although the basic grasp
  of engineering the DNA will come within another eighty
  (80) years or so, if you survive as a species. Our
  friends will still be living when we return home. And
  most of us bring our families with us so we can watch
  our children mature." Their family structure is very
  different from ours and often includes several "models"
  of each set of parents, specifically engineered to
  produce offspring of certain mental character adapted
  to different physical forms needed in their culture.
  All of these live together as a family unit often as
  many as would be in a small village or neighborhood in
  our civilization.

DO NOT DUPLICATE

:TOP SECRET:

* Onboard the "mother ship" almost every race and culture is represented. There is no recognizable form of government behind the actions of these representatives from many worlds. It has been explained that as any species evolves physically, it evolves psychically or spiritually. The outcome of this process is that any form of political organizing, including the politics of religious structure, becomes a "personal moral transgression" of higher shame to any who practice such pursuits, than any good that could come from them. Although this attitude makes any discussion of their faith or beliefs nearly impossible, it has become very clear that most races capable of interstellar travel support the principles of self-determination. In short that there is no empire of conquerors threatening our planet. "A little cross-cultural nudge, like urging a small child to walk towards you, is o.k. But these guys wouldn't let Darth Vader sweep their streets," as a member of a Majestic diplomatic task group put it after several years of analysis of contacts.

Many volumes could be written and have been on the details of this on-going contact process. Much of the contact of individual citizens by these EBE's seems absurd on the surface and we remain largely incapable of controlling such contact. The pattern developing, however, seems to indicate that a course of action is not directly needed. Little threat to national security has occured. In fact, the EBE's have informed us that we are a sort of "protectorate" not unlike a wildlife preserve in their eyes. They will not protect us from ourselves, but are not interested in making us become like them either. There do appear to be forces in outer space that are hostile in the sense that they would exploit us although not conquer us. And we have been informed that a large part of the alien presence on Earth is to isolate us from doing any damage to the alien's own culture until we are more morally evolved as a species. With these things in mind we arrive at a statement of official position.

STATEMENT OF POSITION

"In so far as no threat, either implied nor expressed and either of a military, civil or societal nature has been forwarded by the visitation of the association of visiting extraterrestrials against the governments and the peoples of these United States of America; and in so far as great and many cultural and technical advances have been derived by such exchange, we hereby grant full, complete, diplomatic status and recognition to these individuals from beyond our world. Furthermore, let it be known that we seek to provide shelter, comfort and aid in all their peaceful endeavors in so far as these are respective of the laws of our land and the right to self-determination and free will expressed in our national constitution. And until such time as the objectives or methods of either parties in this agreement shall deem otherwise, this bond between our peoples shall remain in effect."

:TOP SECRET:

EYES ONLY   EYES ONLY

page E-6-of-6

The preceding diplomatic treaty was drafted by the directorate of the MAJESTIC-12 operation and a joint committee of extra-terrestrial visitors and representatives of the U.S. Diplomatic Corps, as a statement of intent. It was ratified and signed at Kirtland Air Force Base, Texas on July the eighteenth, 1954 by President Dwight D. Eisenhower and an individual on the behalf of the EBE's.

Each subsequent holder of the executive office has continued to uphold the intent of this policy towards these aliens.

Beyond the statement of intent, basic agreement has been reached on the following negotiated issues:

- The extraterrestrials will refrain where and when possible from open display of their presence to the public at large.

- The extraterrestrials will submit substantiating proof and listings of all persons contacted from among the public at large or that have been removed from the Earth for purpose of contact or cultural exchange. In no case shall anyone abducted by the EBE's be subjected to knowing harm or kept against their will for longer then forty-eight (48) hours.

- The extraterrestrials will avoid any willful contact with representatives of the public news media or any private investigators of the UFO phenomenon, groups or writers dedicated to the same intentions.

- The U.S. government will provide through the Defense Logistics Agency's Reutilization and Marketing Service such items as needed for the personal comfort of those visiting EBE's who may be sequestered in it's care.

- The U.S. government will provide a section of the MAJESTIC headquarters base in Nevada as an embassy compound for the EBE visitors and equip it to their needs.

- Any exchange of technical, scientific or social information will be conducted item-for-item and in such manner as to assure that both parties have equal gain from the process.

Except for many facinating details which are available under separate cover through either Operation Majestic (MAJOPSHQ) or the Defense Intelligence Agency's Office of Counterintelligence, this concludes the assessment of the situation and statement of position section of this preliminary briefing paper.

EYES ONLY   EYES ONLY

DO NOT DUPLICATE

:TOP SECRET:

S ONLY

EYES ONLY

APPENDIX

AGENCY REPORT NUMBER

405389

EYES ONLY

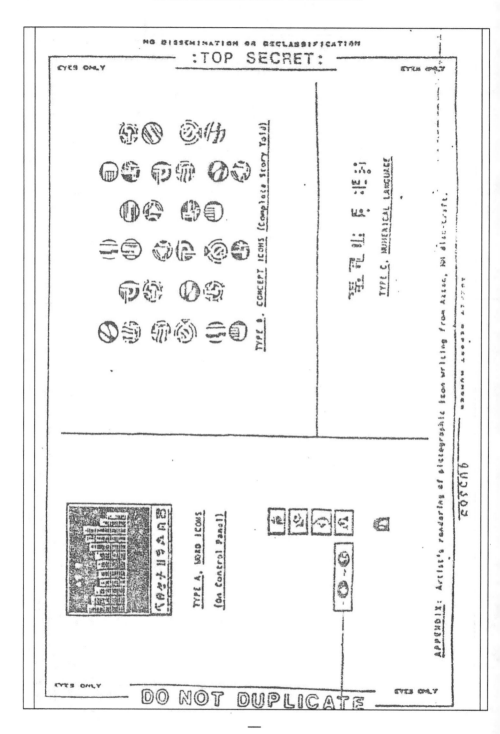

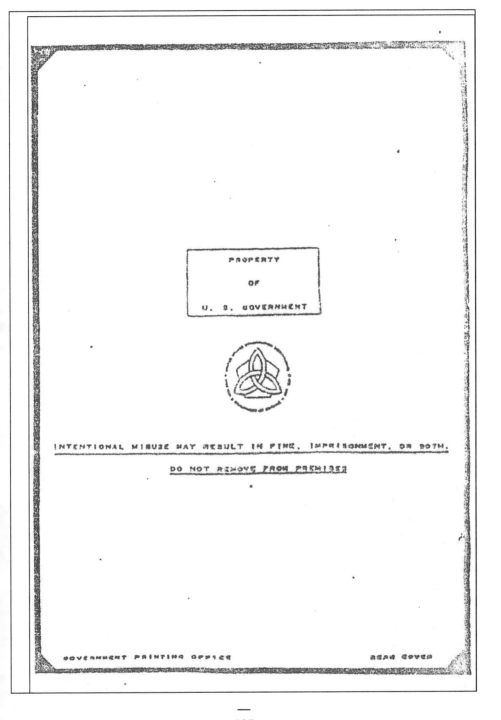